Five-Second Rule

and Other Myths about Germs

The Five-Second Rule
and Other Myths about Germs

What Everyone Should Know about Bacteria,
Viruses, Mold, and Mildew

Anne E. Maczulak, PhD

RUNNING PRESS
PHILADELPHIA • LONDON

Running Press books are available at special discounts for bulk purchases in the United
States by corporations, institutions, and other organizations. For more information, please
contact the Special Markets Department at the Perseus Books Group, 2300 Chestnut Street,
Suite 200, Philadelphia, PA 19103, or call (800) 255-1514, or e-mail
special.markets@perseusbooks.com

9   8   7   6   5   4   3   2
Digit on the right indicates the number of this printing

Library of Congress Cataloging-in-Publication Data is available

ISBN-13: 978-1-56025-950-3
ISBN-10: 1-56025-950-7

Book design by Bettina Wilhelm

This book may be ordered by mail from the publisher.
Please include $2.50 for postage and handling.
But try your bookstore first!

Running Press Book Publishers
2300 Chestnut Street
Philadelphia, PA 19103-4371

Visit us on the web!
www.runningpress.com

CONTENTS

INTRODUCTION

The Microbes in Your Life

The world is full of germs. Every day, we practice good hygiene, cleanliness, and proper social conduct in order to live healthy lives. The germs we live with are more accurately referred to as microorganisms, microbes for short. They are the single-celled life-forms such as bacteria and yeast that live not only in the environment but also on and in your body. Microbes cover every surface in your home and every item you touch at work, on the commute, and at your children's school. They are in the air and in the black ocean depths. They are in food and drinking water. They can remain alive in glaciers for centuries and also in the steamy emissions from deep earth. Is there any doubt that they will soon be recovered from remote sites in the solar system? If, or when, scientists find life on other planets, the odds are good that the life they find will be microbial.

Microbes are the evolutionary precursors to all organisms on earth. By studying their enzymes and DNA, we learn about the biology of humans and other animals. By considering the physical adaptations that microbes make in response to their environment, we gain insight into the ways plants and animals fit into an ecosystem. When observing evolution in a bacterial culture—conveniently rapid due to exponential growth—we answer questions about natural selection.

These same microbes that help us answer our greatest questions about life are so small that they can only be seen under a microscope. Magnifications of 600 are typical. It's no wonder that these powerful beings, visible only with the use of sophisticated equipment, are so overlooked and misunderstood by those of use who deal with them only with our naked eye.

But fear not, you know more about microbiology than you realize. When you start the day by brushing your teeth, using mouthwash, and showering, you are following principles of microbiology. When you rinse an apple, cook your egg, and return a carton of milk to the refrigerator, you are conducting good microbiological techniques. And when you wash your hands, cover your mouth to cough, or wipe down the kitchen counter after preparing a meal, you are a microbiologist at heart.

Still, while microbes influence you in many ways you already know, they affect you in hundreds of more subtle ways. It is within this realm that microbial myths, many containing nuggets of truth, abound. A good example is the "Five-Second Rule," which states—as any child on the schoolyard will tell you—that a cookie dropped to the floor is still okay to eat if you pick it up within five seconds. The "rule" is known more as a yarn than a tenet of biology. And yet this book shows how the Five-Second Rule illustrates six of the basic principles of environmental microbiology, principles that will help you learn the truth about microbes and the tricks for living safely in their world.

This book describes a variety of important microbes: the good, the bad, and the ugly. It highlights the benefits you receive from many of them and discusses those that can become dangerous if not controlled. The goal of *The Five-Second Rule and Other Myths about Germs* is to help you understand how you fit into the natural world, and learn the ways microbes affect health and well-being. And you just might also learn a fun fact or two and gain some helpful tips along the way.

1
What Is Microbiology? A Few Basics

There are only two ways to live your life. One is as though nothing is a miracle. The other is as though everything is a miracle.

—Albert Einstein

They are on your skin right now. You swallow them with your food and drink every day and they're inside you by the trillions. They are in your clothes. And in your bed. You inhale them with every breath. Some are chewing on the pages of this book.

The largest population of living things on earth is the one you cannot see. They are microorganisms, or microbes. Microbes is a collective term for biology's unicellular organisms—they consist of only one cell. They are bacteria, yeasts, and protozoa. Some algae are unicellular. Viruses—which are particles, not living things—are often included in the field known as microbiology. The terms "microbe," "organism," or "microorganism" are used interchangeably by microbiologists. The microbes that make you sick are often referred to as "germs." In certain contexts, the word "bug" is used. If you tell a friend, "I've caught a bug," they will immediately understand that you have a cold or an intestinal ailment and will not presume you have taken to stalking the yard for grasshoppers.

In addition to covering items in your home, hiding in food, and flowing from your tap, hundreds of millions of microbes inhabit your digestive tract. They live on your skin and scalp, they stick to the inside of your mouth. And whether you realize it or not, you are happy to have most of these microbes living with you. They help regulate activities in and on the epidermis. They help digest food and even supply a vitamin or two. They protect against attack from more dangerous microbes.

Unfortunately, the dangerous ones do get through the body's protective barriers from time to time. Sometimes the consequences are merely an annoyance, but in other instances the breakdown of that invisible fortress leads to illness or disease. Mostly, however, your body's normal microbial populations help keep you healthy.

Some microbes live brief lives of twenty minutes or so before they divide and split into two new cells. Some become dormant and return to life after centuries buried in the soil or hidden in a mummy's tomb. Under favorable conditions bacteria multiply and produce several thousand generations within hours. This fast growth, coupled with their ability to move genetic material between cells, helps bacteria evolve quickly to suit their environments. Viruses are different from bacteria, but they have the capacity to mutate and evolve in common. One of the downsides of the ease with which bacteria and viruses evolve is the emergence of resistant mutants that make antibiotics ineffective.

Some microbes possess attributes for surviving in locales that mammals would find lethal. For instance, bacteria have been recovered from the steam vents in Yellowstone National Park. Microbiologists have isolated bacteria from the acid runoff from mines and from the salty waters of the Dead Sea. Some bacteria even have tiny magnets inside, allowing them to continually orient themselves toward the North Pole, perhaps as an aid for direction and movement. Bacteria and fungi carry all the biochemical tools and protective armor they need to live in these and other harsh environments.

Microbes, though tiny, are very resilient. They are infinitely more prepared to live and reproduce in extreme conditions than are higher organisms such as humans. The species that live in close association with humans, in fact, have a rather plush lifestyle compared with those that carry on precarious existences in barren deserts or even in the purified water used in making semiconductors.

Microbiology in the Laboratory

Microbiologists

A microbiologist is a person who has been trained to grow and nurture bacteria and fungi. A mycologist specializes only in fungi, which include multicellular molds and mildew, as well as unicellular yeasts. A bacteriologist specializes in bacteria. Molecular

---

DNA and RNA

DNA (deoxyribonucleic acid) is a large molecule made of two strands of sugars and nitrogen compounds called nucleotides. It also has phosphorus molecules, each surrounded by a few oxygens. The two strands twist into a loose coil and become joined by chemical bonds between the nucleotides. The strands plus the connecting bridges make DNA resemble a ladder that has been turned clockwise at the bottom end and counterclockwise at the top. It holds within its structure all the genes needed by a chimpanzee to be a chimp, a horse to be a horse, an oak tree an oak, or a bullfrog a frog. All living things have DNA, and all living things depend on their unique DNA to contain the instructions for making new offspring of the same species.

RNA (ribonucleic acid) is single stranded. It differs from DNA by having a different sugar backbone and a slightly different set of nucleotides. RNA plays critical roles when DNA replicates inside a cell. DNA replication is vital to any living thing. It is the first step in passing genes from parent to progeny.

---

biologists concentrate on the components inside microbial cells, especially nucleic acids—DNA and RNA.

A microbiologist's job is to determine the conditions in which microbes grow. This information is then used to develop products that either kill dangerous microbes or help the growth of beneficial microbes. Microbiologists in medical fields study the growth of infectious microbes in order to answer questions about disease and to develop effective drugs and therapies.

Because most microbes grow and divide quickly—a dividing time of several minutes to a few hours is not unusual—the microbiologist must keep microbes supplied with a fresh batch of nutrients and regular waste removal.

How to Grow Microbes

A bacterial cell does not get bigger and bigger as it grows until it fills a room. Bacteria grow to a constant size that is characteristic of their species. Their "growth" comes through dividing, growing for a while, and then dividing again. This happens over and over until their numbers reach the billions.

There are two ways to grow microbes in a laboratory. They can be grown on a solid surface in a shallow petri dish containing a layer of agar, a gelatin-like substance, made from seaweed, that has the consistency of very hard Jell-O. Microbes are also grown (or "cultured") in test tubes or flasks containing nutrient-rich liquid called broth (Figure 1.1).

After a few cells of a mature culture (the inoculum) are put onto agar or into broth, the new culture is covered and placed in an incubator. Incubators are simple boxes with shelves and a door, and an electric source for heating. Incubation usually occurs in a range of 72 to 98 degrees Fahrenheit (22 to 37 degrees Celsius), although many specialized microbes can grow at temperatures far outside this range. The microbes associated with the body grow well between 71.6 and 98.6 degrees F.

Figure 1.1. Agar in shallow petri dishes, or liquid broth in tubes, are inoculated with bacteria or molds. (left) Agar is liquid when heated, and turns solid as it cools. *Copyright David B. Fankhauser.* (right) Bacteria inoculated into broth will turn the liquid cloudy during incubation. *Copyright 2006 ATS Labs and Voyageur I.T.*

It takes from one to several days to grow useful amounts of bacteria, yeasts, or molds on agar or in broth. Like pastry chefs peeking into the oven to check on a pie, microbiologists know their cultures are "done" when the few original cells have multiplied so many times that their presence can be seen by the naked eye. On agar, bacteria show up as little dots called colonies. Fungi on agar are recognized as furry masses like those found on stale bread (Figure 1.2). In broth, the millions of new bacterial cells change the color and clarity of the liquid—and very often they smell!

Figure 1.2. Bacteria and molds are invisible before incubation. (left) During incubation, a single bacterial cell will multiply to billions on agar. The result is a visible colony of cells identical to the original cell. *Copyright 2006 ATS Labs and Voyageur I.T.* (right) The *Stachybotrys* mold grows into a fuzzy mass similar to many other molds during incubation. *Copyright courtesy of Aerotech Laboratories, Inc.*

Assuming you don't have an incubator in your house, how do microbes manage to live in cool places like your closets or the garage?

While warm temperatures inside an incubator and the rich supply of nutrients in agar and broth are optimal, microbes can and do grow at temperatures outside their favorite range and with a very limited supply of nutrients. They grow slowly but are hardier when forced to live in harsh conditions such as garden soil or deep within a carpet. In contrast, microbes associated with the skin, mouth, and intestines have comfortable conditions. The human body is warm and offers water and a smorgasbord of vitamins, minerals, sugars, and amino acids—pretty much the same nutrients humans need for good health.

Microbiologists and Their Specialties

Specialty	Job—To study or produce:
Clinical	Pathogen (disease-causing microbe) identification in hospitals
Environmental	Extreme environments; Biofilms; Outdoor and indoor microbes
Subspecialty:	
Marine	Ocean and freshwater microbes
Soil	Soil bacteria and fungi
Water	Drinking water treatment; Sewage treatment
Exobiology	Life on other planets
Bacteria (Bacteriologist)	Bacteria
Virology (Virologist)	Viruses
Food	Bread, beer, cheese, etc.; Preserving foods
Industrial:	
Bioremediation	Microbes for contamination cleanup
Biotechnology	Bioengineered microbes, enzymes, and drugs; Fermentation

Continued on next page >>

The History of Microbiology
The historical advances in microbiology may be classified into three areas: the lens, biological stains, and the microbes.

The Lens
The first lens was used in a microscope in the seventeenth century to observe "very many small living Animals" in rain- and seawater as described by Antoni van Leeuwenhoek. Scientists built upon early observations and developed precise microscopes to make detailed

Specialty	Job—To study or produce:
Consumer products	Disinfectants; Preserving personal care products
Microbial products	Enzymes for detergents, paper bleaching, brewing, meat tenderizers, leather tanning; Vitamins and synthetic compounds; Septic tank treatments
Molecular	Microbial genomes
Morphology	Cell structure
Mycology (Mycologist)	Fungi and mushrooms
Protozoology (Protozoologist)	Protozoa
Pharmaceutical	Vaccines; Antibiotics; Steroids; Digestive aids; Skin and wound medicines
Systematic	Organization and naming of microbes
Taxonomic	Classification of microbes
Wine	Yeasts used in wine making
Research	All of the above
Academic	Microbiology training
Government	National security; Environment; Applied technology; Public Health

sketches of these single-celled beings. Advances in microscopy have reached the point where scientists can study not only cells, but also their internal structures and molecules. From scanning electron microscopy (SEM), which looks at the outside of cells, to transmission electron microscopy (TEM), which creates images of internal structures, few secrets seem to remain in the microbial world (Figure 1.3).

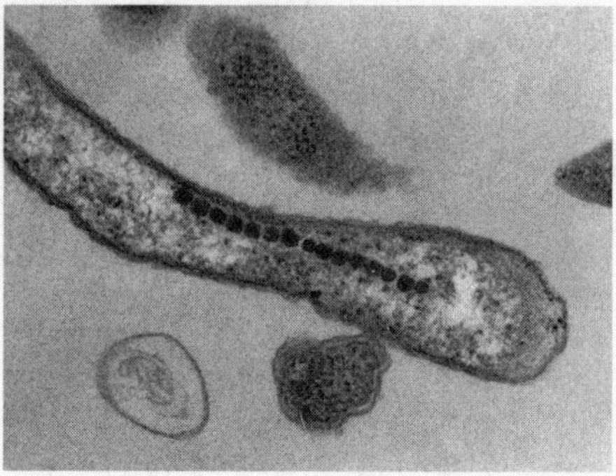

Figure 1.3. Transmission electron microscopy (TEM) is a technique for viewing the cell interior. A string of fifteen magnets is seen inside Aquaspirillum magnetotacticum; magnification x13,535. *Copyright Dennis Kunkel Microscopy, Inc.*

Gram Stain

Microbial staining is low-tech compared with microscopy. By staining specific parts of microbial cells, microbiologists obtain their first clues in identifying them. Without specialized stains, bacteria under a microscope would merely be indistinct specks suspended in a droplet.

The first important staining method remains the cornerstone of identification and disease diagnosis. It is the Gram stain. In 1884 the Danish scientist Hans Christian Gram developed this technique for differentiating bacteria based on their capacity for retaining a purple dye in their cell walls. Those bacteria that capture the stain and look purple under a microscope are called Gram-positive. Other cells cannot be dyed by this method and remain colorless. They are called Gram-negative and are invisible until they are exposed to a second stain, safranin, which turns them pink. Classification of Gram-positive and Gram-negative bacteria is essential in medical, environmental, and industrial microbiology.

Occasionally, you may see the terms Gram-positive or Gram-negative in your community's water-quality report or hear it in the doctor's office. Why is Gram reaction important? Sometimes it provides a warning. For example, numerous Gram-positive bacteria on the skin of a hospital patient are common, but large numbers of Gram-negatives near a wound or inside a catheter are a signal of an impending infection. Microbiologists in hospitals, water-treatment facilities, manufacturing plants, and on food-processing lines all pay very close attention to the presence of Gram-positive and Gram-negative bacteria.

The Germ Theory

The third component of microbiology is the microbial cell itself, the "germ." This seems obvious, yet understanding the concept of "cell" was a major step in unraveling mysteries in biology and the causes of disease. Until the germ theory was accepted, for centuries people had believed that disease happened spontaneously. It was thought to grow out of nonliving objects by chance, and then inexplicably select unsuspecting victims. Some viewed disease more philosophically. They decided that it comes upon an individual as punishment for sin or an evil deed.

Two centuries after the invention of the microscope, Louis Pasteur—most of his studies involved efforts to prevent the spoilage of beer—laid the foundation for modern microbiology by introducing the idea of germs. Among his contributions were:

- Proof that contamination is caused by microbes
- Proof that microbes are transported through the air
- Evidence that they are present in and on nonliving matter
- Evidence that heat destroys microbes

Joseph Lister considered the connection between germs and infection. He was convinced that infections in his surgical lab could be reduced by washing hands and using phenol solution to disinfect instruments. Shortly afterward Robert Koch (pronounced "coke") related microbes to disease. He proposed four steps for linking a specific microbe with the disease it causes. Koch's Postulates captured the principles of good science that are used today. Medical doctors follow these principles in demonstrating cause and effect in disease diagnosis. The postulates, devised between 1884 and 1890, are like most of the central discoveries in the history of science; they are brilliant in their simplicity and clarity.

Koch's Postulates

1. The same microbe must be present in every case of a disease.
2. This microbe must be isolated from a person (a host) who has the disease and must then be grown in a laboratory.
3. If reinoculated into another healthy person or animal, the isolated microbe must cause the same disease.
4. The same microbe must be isolated again, this time from the reinoculated host who has developed the disease.

In general, Koch's principles demand that a disease-causing microbe (1) be present in a patient and (2) its presence is proven by isolating it from the patient. It must (3) re-create the illness if inoculated into another person, and (4) be isolated again to confirm it is the cause of the illness.

Biology has an exception for almost every rule. Since Koch's time new diagnoses have been made. Some microbes follow Koch's Postulates in theory but not to the letter. For example, some viruses and bacteria cannot be cultured by current methods, but they are known to cause disease. As far as we know, disease-causing agents such as the AIDS virus cannot grow anywhere else but in its human and simian hosts. Reinoculating a healthy person to prove Koch's Postulates would be potentially fatal and unethical.

Pasteur's experiments on heating liquids to prevent spoilage led to today's process known as pasteurization, an effective way of preserving milk, beer, juice, and fruit products. If you are curious as to whether Lister whipped up a new mouthwash and gave it his name, the answer is no. Some product names are clever inventions of the marketing department.

Another hallmark in the acceptance of the germ theory was presented in 1928 by the physician Alexander Fleming, who built upon earlier studies on molds. Some of Fleming's bacteria cultures had become contaminated with mold during his experiments. An instant before tossing out the ruined agar plates, he noticed that bacterial growth was inhibited in the area surrounding each mold colony. The mold was *Penicillium* and its by-product that inhibited bacteria was penicillin. What followed was the development of today's antibiotics.

Present microbiology focuses on external and internal structures and how they relate to microbial growth, infection, disease, and disease prevention. Rebecca Lancefield (1934) described entities on the outside of cells (antigens), which led to the field of immunology. Watson, Crick, and Wilkins (1962) are credited with describing the physical structure of DNA, the dawn of molecular biology. Kary Mullis (1993) devised a means for copying and quickly amplifying DNA, a major step that launched today's biotechnology. Environmental science is now using this knowledge to develop microbes for cleaning toxins from land and water.

The Structure of Microbial Cells

Bacteria have all the physical components needed for life packaged in an area not more than about five micrometers in diameter. (A micrometer, or micron, is 0.00003937 inch.) Their shapes vary by species, and each species has its own shape and size that never varies. The distinctive features help microbiologists identify stained bacteria by viewing them under a microscope.

When physicians choose an antibiotic to fight a bacterial infection, the selection of an effective drug is helped immeasurably by knowing the identity (genus and species) of the infectious agent. Viewing bacteria under a microscope is not, however, sufficient to fully identify them. In addition to the physical features, identification of the bug causing your sore throat or infecting a wound requires extra testing. Several simple laboratory tests determine the sugars and amino acids each species can and cannot use for growth. More sophisticated analyses of bacteria's unique DNA, RNA, and fats help confirm genus and species.

Bacteria, protozoa, and all other living things are known by their Latin names, identifying genus (listed first) and species. As an example, the bacterium that causes a certain sexually transmitted disease is named *Neisseria gonorrhoeae*. *Neisseria* is the genus and *gonorrhoeae* is the species. Genus is usually abbreviated to one capital letter; *N. gonorrhoeae*.

Microbial Shape and Structure

Bacteria have several standard shapes (Figure 1.4): (a) coccus, or round; (b) bacillus, or sausage-shaped; (c) vibrio, a curved sausage; (d) spirillum, rigid corkscrew; and (e) spirochete, a long, less rigid, corkscrew. Of the cocci, some species prefer living as single cells, some double up to form diplococci, while streptococci form chains like a string of pearls. The bacteria that cause strep throat are of the *Streptococcus* genus. Therefore, they look like long chains in the microscope. Some cocci grow in bunches such as tetrads (four cells), sarcinae (cubes of eight), and staphylococci (grapelike clusters). "Staph" infections are caused by bacteria of the *Staphylococcus* genus.

The bacterial outer wall has features that allow bacteria to adapt to you and your surroundings. Many bacteria have one or two long tails (flagella) extending off one end. Flagella act as propellers enabling the cell to swim through liquids, such as water, soup, or

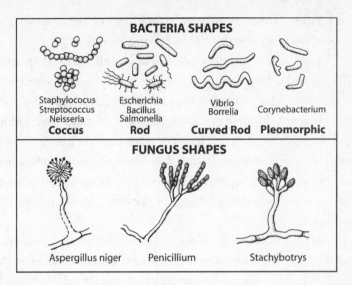

Figure 1.4. The shapes of microbes are characteristic of their species and helpful in identification. *Illustrator, Peter Gaede.*

stomach contents. Vibrio and corkscrew bacteria are excellent swimmers and are referred to as "motile" bacteria. They use a combination of flagella and twisting motions for travel. Examples are *Vibrio cholerae,* which causes cholera, and *Borrelia burgdorferi,* the species that causes Lyme disease. Some bacteria have tiny hairlike projections called fimbriae that help them attach to surfaces. *E. coli,* for

Microbe Terminology

Coccus is singular for a round or oval cell shape; **cocci** is the plural form. **Bacillus** is singular for rod-shaped cells including curved rods, tapered rods, or slender ovals; **bacilli** is the plural form. Staphylococci, streptococci, and bacilli are general terms for any variety of species that are cocci clusters, cocci chains, or rods, respectively.

instance, uses fimbriae to stick to the wall of the intestine, and *Neisseria gonnorrhoeae* attaches to mucous membranes.

The cell wall is complicated compared with the other structures inside bacteria. In addition to holding in all cellular contents, the cell wall keeps out many harmful things. Bacteria depend on their walls for much of their metabolism. While humans have kidneys, livers, and gall bladders, bacteria cell walls perform nutrient uptake, waste removal, energy generation, and "docking" onto plant, animal, and inanimate surfaces. Some microbiologists spend entire careers studying the versatile cell walls of bacteria, found in no other organism on earth.

Another important attribute of bacterial structure is the **spore**. Some bacteria, when under environmental stress, turn into a form (the spore) that has a rugged, impenetrable outer wall. *Clostridium* species—the cause of severe food illness, gangrene, and tetanus—and *Bacillus* are two examples. *Clostridium* and *Bacillus* are not closely related, but because they are able to form spores, they share an ability to withstand extremes in temperature, humidity, and chemicals. *Clostridium* has the added talent of being an "anaerobe," an organism able to thrive in an atmosphere completely lacking in oxygen.

Of the dozens of species in the genus *Bacillus,* the most notorious is *B. anthracis,* the anthrax organism and a potential bioterrorism threat. In a lab, *B. anthracis* cells can be turned into spores by heating them very quickly (Figure 1.5). The resulting spores are then freeze-dried to a very fine white-to-tan powder (the color is from broth the cells were grown in and from powders the spores are mixed into) that may be stored for years.

Anthrax spores are naturally found in soil and have been recovered from ancient sites left undisturbed for centuries. The disease is contracted by inhaling or ingesting the spores, or through open wounds. Anthrax is a rare disease with fewer than two cases reported in the United States per year. In the early 1900s, it was more prevalent

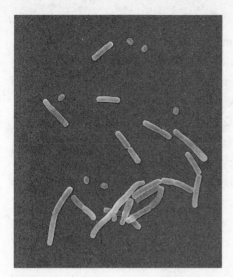

Figure 1.5. Cells and spores of anthrax bacteria, *Bacillus anthracis.* The long rods are growing, replicating cells called the vegetative the form. The dormant, nongrowing spore is smaller and rounder; magnification x700. *Copyright Dennis Kunkel Microscopy, Inc.*

(about 130 cases per year) because people commonly handled livestock hides, which were contaminated with spores from the soil. Of the 236 anthrax cases reported to the Centers for Disease Control and Prevention (CDC) from 1955 to 1999, 153 were associated with processing animal hides or shearing sheep.

Bacteria and viruses are the two major infectious agents in western societies. Protozoa, yeasts, and other fungi play a smaller role. Fungi are ubiquitous houseguests and sometimes go from being annoying pest to health concern. Microbes have favorite places to hide in the home, and they have their favorite ways of disrupting a comfortable routine when left unchecked.

Fungi, Algae, Protozoa, and Viruses

Fungi, algae, protozoa, and viruses are not bacteria. Bacteria belong to a domain of living things called Prokaryotes; fungi, algae, and protozoa are classified as Eukaryotes. Eukaryotes have internal cell structures like those in mammals' cells, and are more complex than the prokaryotic bacteria. By contrast, viruses belong to neither domain and are structurally the simplest of all.

Fungi encompass so diverse a collection of sizes and shapes it seems surprising they could all belong to the same kingdom. Fungi include unicellular yeasts and single mold spores as well as multicellular mushrooms and filamentous molds.

Filamentous molds produce long strings of elongated cells joined together. These filaments can spread great distances and make the mold visible, as evidenced by those growing on bread. One gargantuan mold discovered in 1992 lives outside Crystal Falls, Michigan. The *Armillaria bulbosa* filaments spread in the soil throughout the surrounding county, covering at least thirty-seven acres. As tourists in Crystal Falls fortify themselves on "fungus burgers" and "fungus fudge," they undoubtedly ponder the estimates made on this single mold. The famous *A. bulbosa* weighs at least one hundred tons, or the mass of an adult blue whale, and its age is thought to be at least 1,500 years. Some believe it may be closer to 10,000 years old.

Algae have a diverse collection of physical forms. The shapes of diatoms in particular resemble exquisite and detailed works of art. Two types of unicellular algae prevalent in the environment are green algae and dinoflagellates. Their familiar names are pond scum and plankton, respectively.

Protozoa are the freeform dancers of the microbial world. They do not have a rigid cell wall as the bacteria do, so they change into a variety of shapes as they move through watery environments (Figure 1.6). Amoebae (plural; amoeba is singular) are protozoa commonly studied under the microscope in classrooms because of the fluid and graceful movements they use to seek and envelop food particles.

Viruses are not true microorganisms; they lack

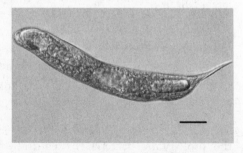

Figure 1.6. The protozoan Euglena is half plant and half animal; it is an animal because it moves under its own power, and it is a plant because it carries out photosynthesis. It is found in freshwater ponds and is not harmful to humans. Bar equals 20 microns. Copyright Jason K. Oyadomori.

sufficient structural components to replicate, divide, and propagate on their own. They must live inside living tissue—plant or animal, or even bacteria.

Viruses are tiny (25–970 nanometers; 1 nanometer = 0.001 micrometer, or one 40-millionth of an inch). They are mere packages of protein-wrapped DNA or RNA in some of the most interesting shapes in nature (Figure 1.7).

The paradox of the virus lies in the fact that its physical simplicity is counterbalanced by an almost diabolical ability to infiltrate living cells and take over the cell's mechanisms as an army might

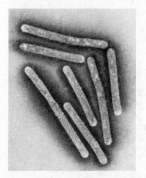 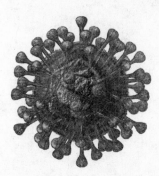

Figure 1.7. (left) Avian influenza is typical of many flu viruses; it made the jump from its animal reservoir of birds, to humans. *Copyright Dennis Kunkel Microscopy, Inc.* (right) Coronavirus, the cause of colds and severe acute respiratory syndrome (SARS). Coronaviruses are covered with clublike projections made of protein, giving them the look of a crown (Latin—corona). *Copyright Russell Kightley Media.*

overthrow the government in a coup. Human cells, though, have formidable defenses to fight off the attack. If a movie were ever made on the offensives and countermeasures between viruses and cells, it would put *Star Wars* to shame.

In these cell wars, viruses attach to specific target cells. (For example, the human immunodeficiency virus, HIV, attaches to T-lymphocytes.) The virus then allows itself to become phagocitized, or eaten, by the host cell. **Phagocytosis** is a normal defensive process wherein a foreign object is drawn into a white blood cell and

Bugs in the News

Microbe	Type	Affliction	Concern
Bacillus anthracis	Bacterial spore	Anthrax	Lethal without treatment
Influenza H5N1	Virus	Bird flu	May infect humans
E. coli O157	Bacteria	Foodborne illness	Diarrhea, vomiting, fever, kidney malfunction
Coronavirus	Virus	SARS (severe acute respiratory syndrome), colds	Potentially Fatal
Drug-resistant *Mycobacterium*	Bacteria	Tuberculosis	Limited drug treatment
MRSA (Methicillin-resistant *Staphylococcus aureus*)	Bacteria	Infections, Septicemia	Limited drug treatment
HIV	Virus	AIDS	No current cure, limited treatment
Stachybotrys	Mold	Contaminated buildings	Respiratory irritation
Salmonella	Bacteria	Foodborne illness	Fever, nausea, diarrhea
Streptococcus pyogenes	Bacteria	Necrotizing fasciitis	Acute infection
Borrelia burgdorferi	Bacteria	Lyme disease	Flulike symptoms and rash
Staphylococcus aureus	Bacteria	Foodborne illness	Nausea, vomiting, abdominal cramping, diarrhea
West Nile Virus	Virus	West Nile Disease	Fever, nausea, Neurological symptoms
Clostridium difficile	Bacteria	Hospital-associated diarrhea	High incidence in children
Hantavirus	Virus	Pulmonary syndrome	Often fatal

Centers for Disease Control and Prevention

then destroyed with enzymes. But rather than being destroyed, viruses use phagocytosis to their advantage. The virus sheds its protein coat once inside the cell and takes over the reproduction apparatus, thus hijacking the cell to make new viruses by the thousands. The host cell becomes a factory for turning out new enemies to penetrate and kill other like cells.

Prions are even simpler than viruses. A prion (pronounced "pree-on") is a single strand of protein that replicates and infects, so rare that scientists long refused to believe such a thing existed. Prions cause mad cow disease (bovine spongiform encephalopathy) and Creutzfeldt-Jakob disease in humans. They accumulate in the central nervous system, eyes, and tonsils. Prions are the most difficult particles to destroy in all of biology; they are able to withstand boiling, freezing, sterilizing, and powerful chemicals. Biotechnology has recently invented an enzyme that inactivates prions, but so far its use has been limited.

Summary

The microbial world is incredibly diverse in sizes, shapes, and lifestyles. These differences are helpful in identifying microbes, an important first step in treating infectious disease. The germ theory was a critical advancement in the understanding of the cell, and present-day microbiology owes its growth to that concept, as well as to the use of biological stains and the perfection of the microscope. Microbes are on almost every surface in your house and on your body. This is the first principle of the Five-Second Rule. If you drop a cookie to the floor, the chances are very good it will meet up with a variety of bacteria and quite a few molds.

2

Microbes in the News

You see, but you do not observe. The distinction is clear.

—Sir Arthur Conan Doyle

We tend to think of microbes as a collection of vaguely mysterious "germs," invisible, sometimes dangerous. Only after they've spoiled the milk or produced unpleasant odors do they lose their mystery. Even then we rarely ponder the various roles they play in our ecology. When microbes make their presence known, they are often viewed as pests.

Humans rarely give these little creatures the respect they are due. Microbes participate in every biological reaction on earth, and had been doing so for more than 3 billion years before humans arrived. They play major roles in recycling carbon, nitrogen, sulfur, and other components of our world, thus replenishing nutrients needed by all higher life-forms. They digest wastes and neutralize environmental toxins. Microbiologist Julie Huber of the Marine Biological Laboratory at Woods Hole, Massachusetts, summed it up by saying, "Microbes drive the planet."

Underappreciated, too, is their sheer quantity. The combined weight of all microbes on earth is about twenty-five times more than that of the total multicelled animal life.

The major microbial groups not only have vastly different shapes and sizes (Figure 2.1), but they also have different types of metabolism. (Metabolism refers to the nutrient usage and biological activities an organism must conduct in order to live.) Knowing a little about each type helps to make better decisions on how to live with the germs around us. For instance, when people don't understand the basic structural and biological differences between bacteria and viruses, they may demand that their doctor prescribe antibiotics, even though they are infected with a flu virus. Antibiotics control bacteria, not viruses. This overuse or misuse of antibiotics in the past fifty years has led to a rise in antibiotic-resistant bacteria. Moreover, symptoms of infectious disease can be tricky to connect to an underlying cause, and even physicians occasionally misdiagnose a disease. Knowing about microbes' lifestyles goes a long way toward identifying the culprit.

Clinical microbiologists and doctors must call upon their knowledge of the overall classification of microbes to accurately identify a pathogen, and then prescribe a drug to kill it. Similarly, a food microbiologist must know the types of microbes likely to spoil a food product in order to formulate an effective preservative system. All microbiology specialists need to know more about microbes than simply thinking of them as generic "germs." It helps you also to understand the various types of microbes and how their differences affect you every day.

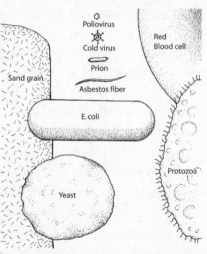

Figure 2.1. Microbes exist in a wide range of sizes. *Illustrator, Peter Gaede.*

Bacteria

Bacteria make news. They cause gastroenteritis and diarrhea (the enteric bacteria), dysentery (*Shigella*), sore throat and ear infection (*Streptococcus*), pneumonia and toxic shock syndrome (*Staphylococcus*), gonorrhea (*Neisseria*), and syphilis (*Treponema*) in addition to hundreds of other diseases. Several types of bacteria in food range from the troublesome to the deadly. The bacteria on your body's surface and those within your digestive tract are harmless unless some sort of disruption puts them in a new situation. An example is *Staphylococcus aureus* (often just called "Staph"), which normally inhabits the nostrils. If these organisms accidentally make their way to a wound on another part of the body, they can cause serious infection that is difficult to treat. Another well-known culprit is *Escherichia coli* (commonly referred to as *E. coli*). Although *E. coli* is a normal resident of the digestive tract, its emergence in other places signals a lack of good hygiene and can lead to illness from body-to-body transfer or from eating contaminated food.

On your body, most of your native microbes have favorite locations: the intestines, the nostrils, the outer ear canal, the scalp, the genitals, feet, skin surfaces, and nails. Others set up shop in and around the home: carpets and drapes, moist bathroom surfaces, drains, tap water, garden soil, pets, and virtually every piece of food you bring home from the grocery. As long as bacteria have water, proper temperature, and nutrients, they can survive for long periods of time in the home. If given optimal temperature, food, and water, they will multiply quickly—just a few hours in the case of food pathogens.

Why E. coli?

Let's get right to the biggest microbial newsmaker. Of all bacteria, *E. coli* is the star. It makes the most headlines. Even before recent *E. coli* outbreaks, it had for decades served as a workhorse in laboratory experiments. By learning about this microorganism, scientists automatically learn a lot about all other bacteria. Despite its notoriety, *E. coli* does not play a crucial role in nature. It resides nowhere else but in human and animal digestive tracts. Within that narrow ecosystem, it is common but outnumbered by many other species.

E. coli doesn't possess unique talents when growing or when initiating infection. It became a favorite in laboratory studies because it grows readily in a petri dish. In this way, bacteria are a lot like people. It's easier to get along with the friendly ones than it is with the high-maintenance types.

It is shaped like a fat sausage—bacillus—and measures about 1 micrometer lengthwise. It's covered with thin projections called fimbrae (Figure 2.2), which help it attach to mucous membranes such as the lining of nasal passages and the intestinal tract. Each cell is covered with hundreds of smaller hairs, pili, which transfer genetic information between bacteria. It is Gram-negative and fairly easy to kill with a disinfectant. It uses oxygen, but in its usual habitat, the intestines, it gets by without any oxygen at all. It does this by switching from an oxygen-using metabolism to alternate means that don't need oxygen for generating energy, which is called anaerobic metabolism. One benefit our bodies get from *E. coli* and other intestinal bacteria is a supply of vitamin K (for blood clotting) and some B vitamins (for energy).

Researchers use *E. coli* in genetic engineering by making it take in DNA from other bacteria to create unique new strains. *E. coli's* own DNA is a circular molecule holding over 2,000 genes. By comparison humans have 20,000 to 25,000 genes. It grows easily on

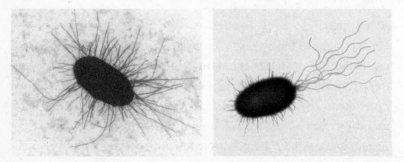

Figure 2.2. (left) *E. coli* is covered with hundreds of fimbriae, which help it attach to the intestinal lining; magnification x6105. (right) The common water bacteria, *Pseudomonas*, use fimbriae and large flagella for movement; magnification x3515. *Copyright Dennis Kunkel Microscopy, Inc.*

water, glucose, salt (sodium chloride), ammonium phosphate (nitrogen source), potassium phosphate, and magnesium sulfate. *E. coli* cells double about every half hour at incubation temperatures around 98.6 degrees F (37 degrees C) so that an overnight incubation of a few hundred cells will give the microbiologist millions by the next morning (Figure 2.3).

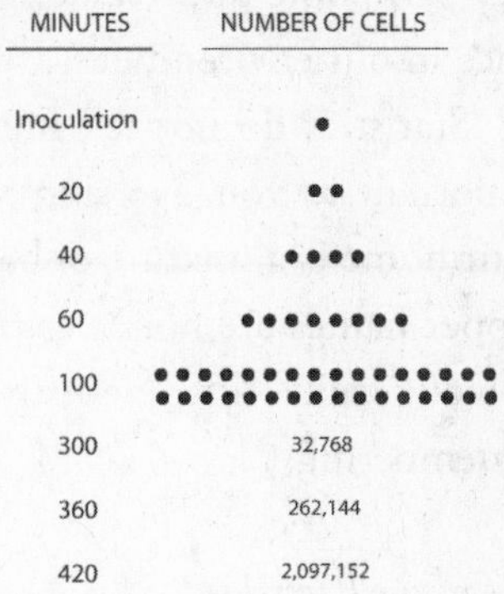

MINUTES	NUMBER OF CELLS
Inoculation	●
20	●●
40	●●●●
60	●●●●●●●●
100	●●●●●●●●●●●●●●●●●●
300	32,768
360	262,144
420	2,097,152

Figure 2.3. If a single bacterium doubles every 20 minutes, after 7 hours there will be 2,097,152 cells. *Illustrator, Peter Gaede.*

All other bacteria possess at least some of *E. coli*'s traits. Its close relatives, the enteric (found in the intestines) bacteria *Salmonella, Serratia,* and *Shigella,* behave similarly. More distant relatives (Gram-negatives) have structures and growth needs like *E. coli*'s, but they possess additional specialties that *E. coli* doesn't have. Examples include *Vibrio,* using its curvy shape to swim through water, and *Thiomargarita,* growing in sulfurous mud and measuring 0.75 mm, visible to the unaided eye. Other species are even less related to *E. coli: Clostridium* forms spores; *Rhizobium* pulls nitrogen from the air for plants; *Chlorobium* carries out photosynthesis.

Microbial Diversity

Scientists do not know how many bacterial species exist on the planet. Estimates begin at six figures and range to trillions. A Princeton geoscientist once estimated over 500,000 different types of bacteria in just a small scoop of soil. Not all microbes on earth have been identified and given names. In fact, microbiologists surmise that they have identified far less than one percent of them. There are 6,000 to 10,000 known species. Environmental scientists continue to find new microbes and learn their traits. By doing this, they gain more understanding about the biological processes taking place in our environment.

Statistics do not tell much about the microbes that are most important to you. No single species sits on your kitchen counter. Your home is a mixture of bacteria, molds, and mold spores. Their proper names are not as critical to know as are the roles they play in your daily life. Nevertheless, there are a few names worth remembering.

Important Bacteria
Salmonella

Salmonella are rod-shaped and motile and are common residents of the digestive tract, especially in cattle and poultry. Pet turtles and reptiles also harbor *Salmonella.* Humans often contract salmonellosis directly from meat and poultry products that are contaminated with various *Salmonella* species, but the microbe occurs from time to time in other foods: milk, ice cream, cream-filled desserts, peanut butter, eggs, seafood, and beef jerky are a few sources of recent outbreaks. *Salmonella* is dangerous when it contaminates kitchen surfaces, water, or food. A toxin (a poison made by a microbe) is released from *Salmonella* cells when they die and break apart in the intestinal lining. The toxin causes the symptoms of salmonellosis: nausea, vomiting, and abdominal cramps, with possible fever, headache, and diarrhea. There are 2 to 4 million cases of salmonellosis in the United States each year.

A particular *Salmonella* species, *S. typhi,* causes an especially serious form of salmonellosis called typhoid fever. In 1906, a feisty, ornery drifter named Mary Mallon took a job as a cook near New York City. Ms. Mallon's hygiene habits were probably questionable at best. Soon, typhoid fever cases in and around the city began to increase. After no small amount of detective work to trace the outbreaks, public health officials found that Mary was the common thread, and, after checking her history for the previous ten years, the officials realized there had been outbreaks in seven of the eight families for whom she had prepared meals. Mary was fired and seemingly disappeared from sight after local officials hounded her into leaving town. She changed her name, moved from town to town, found limited success in odd jobs, then quietly returned to New York years later. A new identity probably helped her find work before long. "Mary" landed a job as the cook in a hospital kitchen. Within three months, typhoid struck twenty-five doctors and nurses, two of whom died. A health inspector studying the cases made the kitchen one of his first stops to trace the source of the illness. Imagine the surprise when the inspector came face-to-face with Mary Mallon. Newspapers dubbed her "Typhoid Mary," adding to her notoriety. Fired again and forced from her home, Mary argued with any and all who would listen that she had nothing to do with people who happened to get sick from her cooking. New York's politicians knew they had a public relations disaster brewing, but were in a quandary as to what fate would be fair for Mary Mallon. After much wrangling, they banished Typhoid Mary to a windblown island in the East River where she lived in relative isolation for the next twenty-three years until her death. One question never answered was, Why didn't Mary get sick from typhoid fever? Typhoid Mary found her place in history as a rare person who carried *S. typhi*, but never herself fell victim to the microbe.

Staphylococcus

The famous member of this genus is *Staphylococcus aureus*, sometimes shortened to "*Staph aureus*" (See Figure 2.4.) It does well in dry environments. Staph is almost always found in the nose. It can be carried in food but is more common contaminating wounds and cuts on the skin. The heavy use of antibiotics in the past several decades has led to an antibiotic-resistant variety. This is MRSA or methicillin-resistant *Staphylococcus aureus*. Despite its name, MRSA is resistant to more antibiotics than just methicillin. It is a particular problem in hospitals, nursing homes, and day care centers.

In 1980, *Staph aureus* caused a new disease—toxic shock syndrome. The microorganisms grew rapidly on absorbent fibers used in certain tampons. The tampons in question were pulled from the market but not before 800 women were affected and forty had died.

E. coli

E. coli is in your digestive tract right this minute, yet your own *E. coli* does not make you sick because your immune system recognizes it as a one of your body's normal inhabitants. *E. coli* from others or from animals can, however, cause severe illness. Though most famous as a foodborne microbe, *E. coli* is implicated in some urinary tract infections and is frequently thought to be the cause of travelers' diarrhea, coming from either food or water.

The *E. coli* species is divided into several specialized substrains that wreak havoc in various ways. They cause illness by (1) releasing toxin that damages the intestinal lining, (2) invading the cells of the intestine, or (3) doing both, invading intestinal epithelial cells *and* using a toxin to create severe damage.

The strain called *E. coli* O157:H7 (O157 for short) has become a current threat to the safety of fresh produce and meats, such as hamburger. (The letters and numbers are designations for structures on the outer surface of the bacteria.) O157 invades the cells lining your intestines. There it releases toxin, creating serious inflammation and bleeding, a condition known as enterohemorrhagic infection.

Streptococcus

Sometimes known simply as strep, these round bacteria are believed to cause a greater variety of illnesses than any other microbe. One common ailment is "strep throat," or sore throat.

Strep is also the cause of dental cavities, tonsillitis, scarlet fever, and rheumatic fever. A recent increase in necrotizing fasciitis is caused by *Streptococcus pyogenes,* also nicknamed the "flesh-eating bacteria." In 1990, a toxic and very fast-growing *Streptococcus* caused the death of Muppet creator Jim Henson.

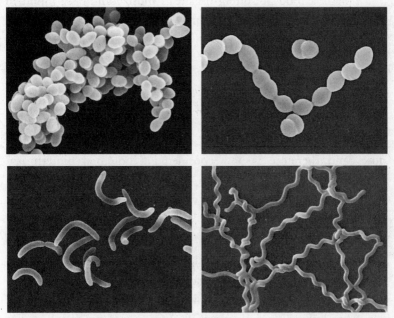

Figure 2.4. Bacteria have distinctive shapes representative of their genus. Shape, Gram stain, DNA composition, and use of sugars and amino acids are used to identify species. (top left) *Staphylococcus aureus;* magnification x3025. (top right) *Streptococcus pneumoniae;* mag. x3750. (bottom left) *Vibrio cholerae;* mag. x2050. (bottom right) *Leptospira interrogans;* mag. x4000. *Copyright Dennis Kunkel Microscopy, Inc.*

Listeria

Listeria monocytogenes often infects soft cheeses and other dairy products. It grows well on foods inside a chilly refrigerator and can live for years in the outdoors, mainly because of its preference for cool temperatures. When ingested, the bacteria enter the body's lymph and blood, resulting in listeriosis. In mild form, listeriosis is characterized by nausea, vomiting, and diarrhea. More serious cases can lead to septicemia (infection in the blood), meningitis, encephalitis, and cervical infections.

Campylobacter

This is a food-poisoning organism most commonly found in meat products. Almost all raw chicken and beef is contaminated with *Campylobacter.*

Shigella

Shigella is a close cousin to *E. coli,* but found only in humans, not other mammals. Its toxin is called the Shiga toxin and it causes severe diarrhea or dysentery. Shiga toxin is hazardous in very small doses and therefore is now considered a potential bioterrorism threat.

Clostridium

This is one of the two major genera (plural of genus) that can morph into an almost indestructible spore when stressed. *Clostridium botulinum* is a deadly microbe often detected in slightly acidic canned foods. The poison is a neurotoxin (toxic compound that attacks nerve cells) excreted from the bacteria. The illness is called botulism and paralysis is its main symptom. When not causing botulism, the neurotoxin has brought happiness to many by being employed as the active ingredient in Botox. Other types of *Clostridium* cause tetanus (*C. tetani*) and gangrene (*C. perfringens*).

Bacillus

Bacillus is the other major spore-forming genus. The cells turn into spores when their environment becomes too hot or too dry, or when they detect dangerous chemicals. Some *Bacillus* species can contaminate foods, where their spores are particularly difficult to destroy. Others have been put to work in agriculture and in the cleanup of toxic Superfund sites, the country's worst hazardous waste sites mandated by Congress for cleanup.

Helicobacter pylori

This bacterium has gained importance since the discovery that it is associated with gastric and duodenal ulcers and stomach cancer. The cells are curved, but often take on a mixture of various shapes (called pleomorphic) within a single culture. They swim through digestive juices using short flagella. *Helicobacter* adapts to the extreme acid conditions in the stomach, which kills most other bacteria. In the stomach, it may initiate ulcers by infiltrating the stomach's protective lining where the cells then reproduce.

Legionella

Like *Helicobacter, Legionella* has found a comfy home in an unlikely place. This organism, which causes the pulmonary diseases legionellosis and Pontiac fever, survives in streams but is also found in the water distribution pipes in hospitals and cruise ships and in moist places like humidifiers and air-conditioning units. The devious *Legionella* lives inside aquatic amoeba cells, making it hard to find in water samples and difficult to treat.

Lactobacillus

Lactobacillus provides many benefits to the food industry. It's used in making sauerkraut, pickles, buttermilk, and yogurt, among other foods. *Lactobacillus* produces lactic acid as it grows. This acid production inhibits other bacteria, thus providing an environment for *Lactobacillus* to thrive in without competition. Acids also help preserve foods.

Mycobacterium

Mycobacterium tuberculosis, the cause of tuberculosis, ingeniously invades the body by hiding for a time within white blood cells, which transport it to the lung, where it does its damage. *Mycobacterium* species have high fat content in their outer wall, giving it two advantages. The fat helps *Mycobacterium* resist antibiotics. It also prevents the cells from drying when exposed to air. Since the microbe is transmitted through the air, it remains virulent as it transits the nasal passages.

Coliforms

"Coliform" is a name given to a group of bacteria rather than one species alone. They are found on plants and in soils, as well as in the digestive tract. Fecal coliforms in animal digestive tracts include *E. coli.* City or county water-quality reports sent to homes report total coliforms. The presence of high numbers of coliforms *may* indicate the presence of other more dangerous bacteria. The Environmental Protection Agency (EPA) mandates that treated drinking water contains zero coliforms per 100 milliliters (about seven tablespoons) of water, but it is not uncommon to have small numbers (less than 1 percent of the total bacteria detected) in water on occasion. In recreational waters, there should be no more than 199 fecal coliforms per 100 milliliters. If the numbers are higher, rethink your plans to swim at that beach.

Bacterial Specialties

Bacteria have numerous ways of adapting for survival in unusual circumstances or locations, and because of fast generation times, bacteria can quickly evolve in response to their surroundings. Yet, bacteria and other microbes do not specifically set out to injure humans. Sometimes their adaptations make it possible for them to coexist with human or animal hosts (Table 2.1). Still, many adaptations do promote serious conditions. For example, some *Staphylococcus* species produce coagulase, an enzyme that coagulates blood. Should a coagulase-producer colonize a cut or wound, the clotted blood forms a shield from the body's immune system, allowing the *Staphylococcus* to inflict its damage without resistance.

Fungi and Yeast

There can be up to a million fungus cells, spores, and colonies per gram of soil (smaller than a sugar cube), and their mass far outweighs that of bacteria. They help degrade the world's organic waste, and without them, we would be neck-deep in the stuff. Different types of fungi can be made up of many cells (multicellular) or only one cell (unicellular). (Fungus is singular, pronounced with a hard "g." Fungi is plural, pronounced "fun-ji.") Yeasts are fungi that are found only in a unicellular form. When a new yeast cell forms, the process is called budding, and the new cell is a bud. The fuzzy masses of mold and mildew that are visible to the naked eye are multicellular, but many release single-celled spores into the air as part of their reproduction. The microscopic fungal spores are as important to you as the masses of bread mold and bathroom mildew because they cause infections and allergies.

Fungus cells have many of the same characteristics as human cells; they are both eukaryotes, cells with defined internal structures. Fungal diseases are sometimes more difficult to treat than bacterial diseases because of the similarity between fungi and mammalian cells. Often medicine that kills fungi may unfortunately harm certain cells in the human body as well.

Household fungi can sometimes make your peek into the refrigerator feel like a perilous safari. Fungi do well in cool conditions inside refrigerators or in garden soil, and make the most out of small amounts of moisture. They absorb nutrients from soil, saltwater and freshwater, or from animal or human skin.

Name of Bacteria	Adaptation	Its Specialty
E. coli, S. aureus, C. tetani	Plasmid, a small piece of circular DNA	Carries genes for drug resistance and toxin production
Clostridium and *Bacillus*	Spore form	Protects against heat, cold, and chemicals
Corynebacterium (causes acne)	Many different shapes (pleomorphic)	A structural anomaly rather than an adaptation; purpose unknown
Treponema (causes syphilis)	Corkscrew shape	Swimming
Proteus (causes urinary tract infections)	Many flagella (peritrichous)	Rapid, swarming movement
Aquaspirillum magnetotacticum	Contains magnets made of iron oxide	Movement or protection against oxidizing compounds

Table 2.1. Examples of bacterial adaptations.

Molds grow in filamentous strings with outcroppings that contain packets of spores, which travel great distances through the air when released. (Mold spores are for reproduction and are not related to bacterial spores of *Bacillus* or *Clostridium*. The terminology can be confusing.) When a fungal spore settles on a surface and finds enough nutrients to grow on and favorable humidity, it begins growing into the multicellular monster you are used to seeing. This growth can take place very quickly with massive amounts of new spores released into the air.

Name of Bacteria	Adaptation	Its Specialty
Pseudomonas	Slime or biofilm layer (glycocalyx)	Adherence to surfaces in flowing liquids or water
Streptococcus mutans	Dextransucrase enzyme (causes tooth decay)	Growth on tooth enamel
Thiomargarita namibiensis	Size (sulfur-oxidizing bacteria)	Giant vacuole allows cell to "hold its breath" if nutrients are scarce
Halobacterium	Lives in very high salt conditions	Growth in Great Salt Lake
Thiobacillus	Lives in acid conditions	Growth on drainage from mines
Pyrococcus	Lives at high temperatures	Growth in hot springs over 212 degrees F (100 degrees C)
The barophiles	Live under high pressure	Growth in ocean depths

You are less aware of the yeasts in your life. Some are normal residents on your body and are a problem only if their numbers increase to create an infection. Many other yeasts are used in food production. Though there are not many live yeasts in the foods you eat, it is almost impossible to construct a diet that doesn't contain some yeast-produced foods: bread, baked goods, wine, beer, and vinegar salad dressings.

Killing molds with disinfectants is more difficult than killing bacteria. Yeasts are easier to eliminate with disinfectants but, like molds, their infections in the body are hard to cure.

Studying fungi is part science and part art. Despite the scientific advances used in identifying bacteria, most bacteria remain anonymous. Fungi are an even bigger unknown. There are few tests available for identifying fungal species. Study of environmental molds is therefore a challenge, and diagnosing some fungal infections can be very difficult and cause delays in their treatment. Mycologists use many of the same techniques that have been around a hundred years; they look at a mold under a low-powered microscope. Surprisingly, a talented mycologist becomes quite skilled at recognizing molds and their distinctive spores this way.

Important Fungi

Penicillium

As the superstar mold, *Penicillium* produces penicillin. Shortly after its discovery, penicillin played a pivotal role in World War II that may have expedited the war's ending. Infections in U.S. soldiers wounded on the battlefield were treated with the new "miracle drug," while Axis troops, lacking a supply, suffered more fatalities from wound infections. Penicillin is therefore thought of as one of several critical factors that altered the course of history during the 1940s.

Penicillium grows at room temperature and slightly cooler and has an appetite for a wide variety of foods, including bread and fruit. Its visible colonies range from blue-green to grayish to yellow-green.

Aspergillus

This close cousin to *Penicillium* is used for making soy sauce, but can be a nuisance when it contaminates fruits and vegetables and peanuts. *Aspergillus* mold spores drift through the air with the breeze, and when large doses are inhaled, the spores cause mild-to-serious respiratory infections. Compost piles can contain large amounts of *Aspergillus* spores, so gardeners often suffer the consequences when they inhale them. *Aspergillus flavus* produces a toxin called aflatoxin that, when ingested, has been linked to cancer of the liver. Aflatoxins are serious health hazards in many parts of the world where heavy mold contamination on crops is common. When working in dusty buildings, barns, or places near dry hay or grain, potentially high amounts of airborne spores should be avoided by wearing a mask that covers both the nose and mouth.

In households *Aspergillus niger* (pronounced "A. ni-jer") is recognizable by its black color on bathroom tubs, tile, and tile grout.

Candida

Candida albicans is a yeast. It causes candidiasis in women and thrush, an oral infection, in adults and children. *Candida* (pronounced CAN-de-da) is found on your skin where your normal assortment of skin bacteria keeps yeast numbers under control. When taking antibiotics for a bacterial infection, the reduction in bacteria allows *Candida* (a fungus, so unaffected by antibiotics) numbers to increase. It will overrun mucosal tissue, and the result is an annoying and pesky infection.

Trichophyton

Think "athlete's foot." This mold has a genius for invading the human body and staying there. Like many dermatophytes (fungi that invade skin, hair, and nails) it is difficult to eliminate. Preventive measures work better. For example, wear sandals in the locker room, even while showering.

Stachybotrys

Stachybotrys mold, or "Stachy" (pronounced "stacky"), becomes a newsmaker shortly after a community has suffered flood damage. Its black growth spreads across water-damaged surfaces and inside moist walls. Health complaints are often associated with

Stachybotrys-infected buildings. Stachy has been nicknamed the "black mold" or "toxic mold." It is believed to release into the air a mycotoxin (any toxic compound produced by a mold) that causes severe respiratory irritation and allergy. Although health threats from *Stachybotrys* are not thought to be increasing, litigation regarding "mold-infected" buildings is on the rise. Note that buildings cannot be infected as people are infected. It is more correct to say *Stachybotrys* causes mold *damage* to buildings.

Mildew

What is mildew? It's a general term to describe the visible layer of mold growing on damp surfaces. Leather, paper, fabric, and fruits and plants are most vulnerable to mildew, which can stain them or cause permanent damage. *Aspergillus* and *Penicillium* are the common household molds that also double as mildew.

Almost all the molds in your home are from the outdoors, and at their densest in the summer and fall. More mold spores have been detected in the outside air of the U.S. southwest, far west, and southeast, compared with the north, northeast, and New England. Though certain fungi are damaging or are health hazards, fungi bring many benefits. In addition to eating away the earth's waste materials and recycling nutrients, fungi and yeast help in making numerous foods, as well as drugs and industrial enzymes.

Protozoa and Algae

Protozoa, amoebae, and algae are often used as specimens for study in classrooms. Their movements are fun to observe under a microscope, and their cells contain defined organelles that bacteria lack. (An organelle is a small specialized part of a eukaryotic cell. For instance, the nucleus is an organelle.) Amoebae constantly change shape as they slink through their watery surroundings.

Other protozoa may be more uniform in shape than the amoebae, but they, too, can squeeze and elongate, if needed, to get to where they're going.

Protozoa are not found on household surfaces. A common indoors location is inside fish tanks. Several species of protozoa live in the intestines without causing trouble. They contribute to food digestion but probably to a lesser degree than the intestinal bacteria.

Protozoal infections are more common in geographic areas with poor sanitary conditions or high concentrations of insects that carry certain protozoa. Infections are often the result of drinking untreated water. Some of the protozoa that may be ingested with untreated or partially treated water are *Giardia, Cryptosporidium, Entamoeba histolytica, Cyclospora, Balantidium,* and *Naegleria.* The insects that are implicated in carrying specific protozoal species that cause human infections are sand fly (carries *Leishmania*), tsetse fly (*Trypanosoma*), and *Anopheles* mosquito (*Plasmodium,* the cause of malaria).

Giardia and Cryptosporidium

Protozoal diseases are comparatively rare in North America versus other parts of the world. There are two protozoal parasites, however, that are frequent in the United States: *Giardia* and *Cryptosporidium.* Both are ingested by drinking unfiltered water from streams, and will cause severe intestinal distress that persists for weeks. The protozoa are shed in the feces of wild animals and domestic cattle. Beavers have been most closely associated with *Giardia.* *Cryptosporidium* has been found in the wastes from numerous animals, but cattle may be its most important source because of the trend for communities to expand near farming areas.

Giardia is difficult to diagnose, but when diagnosed, the infection is treated with the drugs metranidazole or quinacrine hydrochloride. Giardia cells attach tightly to the intestinal lining, and if left untreated, they spread enough to interfere with nutrient absorption. The symptoms of giardiasis are malaise and weakness, nausea, gas, abdominal cramps, and weight loss.

Cryptosporidium, or "Crypto" (pronounced "krip-toe"), forms a very resistant cyst during part of its life cycle in order to withstand environmental stresses. Crypto is resistant also to chlorine. It must be in contact with chlorine disinfectants for a long time (hours compared with minutes) before it is destroyed. In 1993, *Crypto* made news when it caused sickness in more than 400,000 people and one hundred deaths during an outbreak in Milwaukee. The outbreak occurred after a period of heavy rains overran the capacity of the local treatment plants, leading to inadequate filtering of the community's drinking water.

Both *Giardia* and *Cryptosporidium* are large microbes (*Crypto* is about 7 microns across, *Giardia,* 10–12 microns). Filtration is more effective than chlorination for removing them from water. If you must rely on drinking directly from a woodland stream while hiking or exploring wilderness, use a certified filtration device sold at camping supply stores.

Algae

Algae, like molds, are identified by physical features. Algae are also like molds in that there are multicellular species and unicellular species. The algae commonly found in fish tanks and birdbaths are unicellular green algae and blue/green algae. To make matters more confusing, blue/green algae's name is misleading. Blue/green algae are actually specialized bacteria that absorb nitrogen and produce oxygen through photosynthesis. The proper name for blue/green algae is *Cyanobacteria.*

Green algae grow in water exposed to light. Glass or clear plastic containers are as prone to contamination as fish tanks and birdbaths. Green algae are very hardy and can survive even if the vessel is left to dry out for weeks. When water is added again, the green algae bloom. They may be harmful to fish and birds if they produce a toxin. There are products to eliminate them, if only temporarily. A few copper pennies in a birdbath do the trick and don't harm the birds. Algae-eating fish are often put in fish tanks to help keep the green color under control. Frequent water-changing helps reduce nutrient levels and holds algae growth to a minimum.

If you swim in the ocean, you are immersed in a soup of tiny algal creatures. They are the **plankton** and the **diatoms**. Both have rigid cell walls. Plankton's strength comes from cellulose imbedded in its wall. Diatoms have a hard silica outer wall. In addition to the ocean, diatoms are found on rocks, plants, and in neutral-basic soils in dry or high-limestone areas, that is, where lilacs and herbs grow. Like snowflakes, diatoms form some of the more intricate and beautiful shapes found in nature (Figure 2.5).

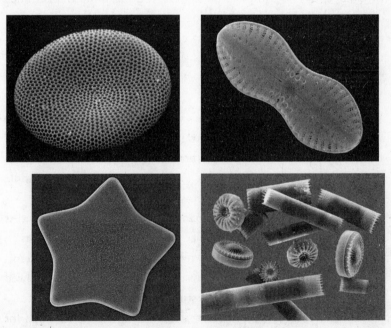

Figure 2.5. Diatoms are algae and a type of plankton that are free-floating in marine and freshwater. Their distinct architecture is made from two complementary pieces that fit together like lock and key. (top left) 121 microns. (top right) 45 x 13 microns. *Copyright Kenneth M. Bart.* (bottom left) Magnification x345. (bottom right) Mag. x100. *Copyright Dennis Kunkel Microscopy, Inc.*

An important diatom is *Pseudonitschia pungens,* producer of domoic acid. This compound has caused deaths in people who have eaten mussels that feed on *Pseudonitschia.* Domoic acid poisoning is currently causing an alarming loss of life among marine birds and sea lions on the west coast.

Viruses

Viruses are very simple structures. They lack much of the infrastructure found in bacteria and protozoa. Viruses have evolved to a form that lacks a complete reproductive apparatus that would enable them to multiply on their own. In order to reproduce, viruses must infect a plant or animal host. (Some viruses called bacteriophages infect bacteria.) That is why viruses are sometimes thought of as the "ultimate parasite." Nothing more than strands of DNA or RNA wrapped in protein, viruses shed even their protein coat when they enter a host cell. Once inside the host, they retain just enough power to unleash trouble.

Virus Basics

Viruses are not found free in nature as are bacteria and protozoa. The ultimate parasites can more accurately be called obligate parasites, meaning they must depend on other living things to survive. When the host sheds viruses into water or food, or onto a faucet handle (among many other inanimate items), they can cause infection.

Many viruses that infect humans start in an animal reservoir. As bird flu and swine flu have shown, viruses jump from reservoirs to susceptible human hosts. This is called a species jump. Upper respiratory tract infections are the most common viral ailment, followed by those that affect the gastrointestinal tract.

Some viruses remain infective while sitting out in the open for a period of time. An example is rotavirus, which is known to be transmitted in day care centers. Rotavirus is picked up by children from toys and from the floor in addition to contaminated food. Other virus types don't last long outside the body, such as human immunodeficiency virus, or HIV, the cause of AIDS. Most infective viruses that accidentally end up on a bathroom tile or kitchen counter are easy to kill with disinfectants. A few, such as the cold virus and hepatitis A, are slightly more resilient. As you will see, following the directions on the disinfectant can is important when you suspect there are viruses nearby.

Viruses range from being benign pests (certain papillomaviruses cause warts) to dangerous agents (human papillomavirus is the main cause of cervical cancer). Some, like Ebola, are deadly. Latent viruses are troublesome because they hide in specific body tissue for a very long time. Because of **latency**, it's difficult to deduce where or when you may have caught a particular virus. It may have been months, years, or even decades ago. Species belonging to the herpes viruses hide inside nerves for long periods before causing shingles, sexually transmitted infection, canker sores, skin infections, or serious neurological disease.

Virus Fast Facts

- Viruses are fifty times smaller than bacteria; 2,000 times smaller than animal cells; one sixteen-thousandth the size of the head of a pin. Scanning electron microscopy and X-ray crystallography are special techniques used for viewing the artful submicroscopic world of viruses.
- The incidence of colds from a specific virus is nearly impossible to calculate. There are over one hundred rhinoviruses and the immune system seems overmatched in defending against assault from their infinite combinations. In addition, at least 30 percent of colds are caused by coronavirus, distinct from rhinovirus. We pick up cold viruses by touching objects contaminated by others, then inoculating ourselves with the virus

by touching hands to the eyes, nose, or mouth. Cold viruses also arrive at your face when an infected person thoughtlessly sneezes without covering up. Once the virus has located a mucous membrane to invade, it is inside the cell and within a half hour busy making new cold viruses by the thousands.

• Virus shapes are varied and often geometric (Figure 2.6). Rabies virus, for instance, has a helical structure. That is, its RNA is coiled inside a cylinder. Several others are polyhedral, composed of twenty triangular faces and twelve corners. Rhinovirus, herpes, hepatitis B, and rotavirus are polyhedral. Bacteriophages have a complex shape that looks a bit like a lunar landing craft.

• All living things have their own group of viruses that attack them. There are viruses that specifically target humans, other animals, bacteria, fungi, algae, plants, insects, and so on. But many (flu, HIV) can jump to humans from other species.

• Within their favorite species most viruses prefer a specific cell type. HIV attacks T-lymphocytes in the blood; papillomavirus seeks only skin cells; hepatitis B goes right for the liver.

• Rotavirus is the most common cause of diarrhea in children; 55,000 infected children are hospitalized each year in the United States due to rotavirus.

• Norovirus (a collective term for a group of similar viruses called Norwalk viruses) is estimated to be responsible for half of all gastroenteritis

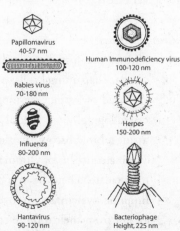

Figure 2.6. Viruses are classified according to shape, outer coat, and whether they contain DNA or RNA. *Illustrator, Peter Gaede.*

(nausea, diarrhea, weakness, sometimes with fever and chills) cases. It is often referred to as "stomach flu."

Disease Terminology

Disease—A change in the healthy state of the body resulting in the inability to carry out all normal functions. **Infectious disease**—Disease caused by a microbe capable of invading the body and remaining in the body to carry out part or all of its activities. **Communicable disease**—Disease that can be transmitted from person to person. **Contagious disease**—Disease that is *easily* transmitted from person to person. **Infection**—The invasion and replication of a microbe in or on a host. **Infectious**—Type of microbe or dose in microbe numbers that can cause an infection.

Summary

Our world holds a wide variety of microbes of almost infinite diversity, and they are found almost everywhere on the planet. Bacteria cling to biological and nonbiological, or inanimate, surfaces and most have a certain type of habitat they prefer. Mold spores are carried on the breeze and settle on almost every surface in your home. Most bacteria and molds live independently; they can exist outside the body. Viruses have no choice but to infect a plant or animal host cell. If they didn't, they would rapidly disappear from the earth. Protozoa are somewhere between the bacteria and viruses in how finicky they are about their surroundings. They require liquid environments, usually ponds and streams. Trouble starts when the wrong microbes— bacteria, viruses, or protozoa—find their way into your body.

3

We All Live in a Microbial World

What a revelation any science is!

—Howard T. Ricketts

You probably go about your daily routine in the same manner as a microbiologist. Yet microbiologists may differ from you in one important way. They notice the microscopic world that influences every touch and every breath. If *you* are observant, you also may become pretty good at "seeing" germs, despite their invisibility to the naked eye.

When you wake up and brush your teeth, you are reducing the numbers of cavity-causing and odor-producing anaerobic bacteria in your mouth. A quick shower rids you of a few billion bacteria and yeasts and helps decrease odors. While showering you may notice that strange pink color on the tile grout. Preparing breakfast, you cook eggs, fry ham, and replace the milk carton in the refrigerator. At work, you may have decided it's a good idea to wash your hands after using the restroom. And you have no doubt cast a wary eye toward the kitchen in that fast-food joint your coworkers love so much. All these activities have a basis in microbiology.

Microbes at Home

All household microbes have a common bond—their need for water. Moisture helps viruses remain infective for periods when they are without a host. Bacterial spores that have been dormant for years awaken and grow in the presence of water. Molds get by on less moisture than other microbes, but they, too, need some. Water is also a vehicle for spreading disease-causing microbes: aerosol droplets from a sneeze, contaminated drinking water, spoiled potato salad, and so on. Always keep in mind that moist environments are the first spot to consider when looking for molds and bacteria around the house.

Bathrooms

Most people assume the bathroom is the most likely place to find dangerous microbes. The makers of cleaning solutions have used the bathroom for years as an example of Germ Central, educating consumers through television ads on the hazards of dirty sinks, tubs, and commodes. Indeed, scientists have found *E. coli* on the inner and outer surfaces of toilet bowls, on the flush handle, and on the floor. Flushing the toilet while the lid is open may release bacteria-laden aerosols that can travel up to twenty feet, easy striking distance to other bathroom surfaces, even the toothbrush! But there is good news. Though fecal organisms are found everywhere, their numbers on sink countertops, the toilet seat, and the floor are low. Only forty-nine or so bacteria per square inch are found on toilet seats. The largest amounts of microbes are in the shower drain and on the toilet flush handle.

Except fecal microbes, most bathroom nasties are more nuisance than health concern if you clean regularly. But almost any microbe can transform from benign to dangerous if the conditions are right. For instance, the pink stuff that grows in the shower stall, on bathroom fixtures, and at the toilet bowl waterline is *Serratia marcescens* bacteria. *Serratia* is a normally harmless microbe found in water. If it infects the body, it can turn dangerous, perhaps aided by the rich supply of nutrients it finds in bathroom dirt. When given an opportunity, *Serratia* infects the urinary tract, wounds, or the lungs, causing pneumonia.

Two household microbes found in bathrooms are the molds *Aspergillus* and *Mucor*. They are the black and mossy green mildew, respectively, that appears on bathroom tile grout. They, too, may cause health concerns when large amounts of their spores are inhaled. *Aspergillus* is troublesome outside bathrooms, as well. Spending long periods of time in a very dusty cellar with little ventilation increases the chance of inhaling *Aspergillus* spores.

Kitchens

Studies on kitchens and bathrooms sample thousands of homes to find how people manage microbes in a "typical" American household. They tend to understand the need for cleaning the bathroom, and most toilets and tile are washed regularly with chemical cleaners and disinfectants. Yet the message is often lost when taking the short walk to the kitchen, where the battle against germs is waged with an ancient sponge and a few drops of tepid water.

Microbiologically, the kitchen is the filthiest place in your home. The kitchens that look spotless are frequently found to contain more microbes than kitchens that are grungy. There are several reasons, some obvious and some subtle.

The kitchen is a high-traffic area for all family members and often has direct access to the outside world. It is the entry destination, storage point, and preparation area for a huge source of microbes that are both dangerous and harmless. That source is food. And since many people don't want to use the same strong chemicals near their food that they use to clean the bathroom, they often spread more microbes around the kitchen than they remove. An old wet sponge or rag doesn't kill microbes, it just moves them around. That's why the "clean" kitchens may actually have enormous amounts of bacteria and molds, and maybe a few viruses, on every surface.

The most microbe-contaminated sites in the kitchen are sponges and dishcloths, the sink drain, faucet handles, and cutting boards. Environmental microbiologist Charles Gerba, PhD, has reported that there are "two hundred times more fecal bacteria on the average cutting board in the home than [on] the toilet seat."

Cutting Boards

Meats and vegetables are washed and handled at sinks and on cutting boards, so these places can be loaded with potential pathogens. Washing a sink and rinsing it well with hot water gets rid of a good amount of these dangerous microbes. Cutting boards are the challenge. Plastic cutting boards retain bacteria, especially if the plastic is heavily etched with knife cuts. The deep slashes give microbes a good hiding place, especially if fatty materials enter the cracks and provide a moist and protective barrier for the microbes.

The porous nature of wood enables wood cutting boards to absorb microbe-laden liquids more easily than do plastic boards. Wood cutting boards have been shown to retain live bacteria up to twelve hours after being exposed to chicken juices. Eventually, the bacterial numbers within a contaminated wood cutting board or in the crevices of plastic boards decline within hours. Microbiologists have shown that the type of wood—ash, basswood, beech, birch, cherry, maple, oak, or American black walnut—does not affect its capacity to retain microbes.

Sink Drains

There's no surprise that the kitchen sink drain has high concentrations of bacteria. It offers plenty of food and water and an environment that occasionally washes away microbial wastes. When sinks are left unused for several days, a specialized group of bacteria grow in the stagnant conditions. After a vacation the kitchen drain may give off a rotten-egg smell or other bad odors. The odors are by-

Tips for Cutting Board Care
- Discard plastic or wood cutting boards that are excessively scratched.
- Scrape off food immediately after using boards and before washing. Wash boards with hot soapy water and rinse well immediately after use.
- After washing, pat the surface dry with paper towels, then allow to dry completely. Air drying alone may not be effective if the board material readily absorbs liquids.
- Sanitizing the board may not be necessary if it is cleaned immediately after use, but if sanitizing, use dilute bleach solution (one teaspoon to one quart water) and soak for three to ten minutes.
- Clean dishwasher-safe cutting boards in the dishwasher.
- Do not chop salad vegetables, fruits, or put any other ready-to-eat foods on a cutting board that has just been used for raw meats or fish. Always clean in between uses or keep separate cutting boards for each type of food.

products of microbial growth in the low-oxygen, high-nutrient drain water. Running the water for a few minutes is the cure. Pouring a small amount of bleach into the drain helps, too.

Sponges

Kitchen-cleaning causes more problems than it solves if done incorrectly. Soiled kitchen sponges, cloths, and mops are so laden with microbes that using them for wiping surfaces after a meal merely spreads the contamination. Sixty-seven percent of sponges from households have been found to be contaminated with fecal coliforms, their numbers in the neighborhood of 30 million in a 2 x 2–inch section of sponge. The bacteria commonly found all around the kitchen, in addition to coliforms and fecal coliforms, are *E. coli*, *Staph*, and *Pseudomonas*. *Salmonella* and *Campylobacter*, both causes of foodborne illness, also show up often. Raw vegetables, meat, and fish are sources of all these bacteria, except for the ubiquitous water microbe *Pseudomonas*. Therefore, diligent after-dinner wipe downs with used sponges or dirty towels make the kitchen a more microbe-laden place than if you had never cleaned it at all.

A clean-looking sponge may hold hundreds of bacteria. An obviously dirty sponge—one that is discolored, smells bad, or you *know* has just wiped up raw meat juices—surely has thousands to millions of bacteria in it. A few methods for ensuring the cleanest sponge possible are: (1) soak for up to five minutes in bleach diluted two and a half teaspoons to a cup of water; (2) microwave moist sponges for one to three minutes and let dry; or (3) throw dirty sponges in the garbage and replace with new ones. Other good ways to reduce germs in sponges are cleaning in a dishwasher; cleaning in the washing machine; boiling for five minutes; or soaking for five minutes in 70 percent isopropyl alcohol or distilled white vinegar. After soaking, rinse the sponge in water before using it.

Hot, soapy water and a *clean* sponge are excellent tools for kitchen cleaning. Always follow by rinsing surfaces with water. In a circumstance in which you fear that raw meat juices may have splattered, follow the soap and water with a disinfectant or sanitizer. After following the product's directions, rinse or wipe down the surface well to remove residual chemicals.

Viruses are not common in the kitchen unless a family member with a viral infection sneezes or coughs or touches contaminated skin to a surface. Cold and flu viruses survive up to three days on such surfaces as counters, tables, and the outside of a refrigerator. Since fecal bacteria are known to find their way into the kitchen, it would be no surprise to also find an occasional enteric (from the digestive tract) virus lurking near the dinner plates. If you're looking for a conversation topic at your next party, mention that it's safer to eat the hors d'oeuvres in the bathroom than it is in the kitchen.

Laundry

You've worn your clothes all day and they're making you feel dirty. Throwing them and some detergent in the washer should help, right? Think again. Microbiologists have found *E. coli* in samples from wash cycles and in washed clothes. Regular detergents do little to kill microbes. Up to one-fifth of washers contain *E. coli* and 25 percent have fecal matter. Washing underwear with other clothes or with dish towels simply spreads the "wealth." If an item doesn't have fecal microbes before washing, it likely will after the wash cycle.

Hot water cycles and laundry additives that sanitize both help reduce contaminants in washing machine loads. To save electricity costs, however, many people don't use hot water washes, and only a small percentage use bleach or laundry detergents labeled as sanitizers. So while washing with regular detergent removes dirt and grime, it leaves behind a vast assortment of bacteria and even some viruses.

The dryer kills some of the less hardy microbes, but only a few. *Salmonella* bacteria and hepatitis A virus can survive a twenty-minute wash and a twenty-eight-minute drying. In other words, your clothes might come out of the dryer "dirtier" than they went into the washer.

Carpets and Walls
Mold Hazards

Carpets and floors get heavy foot traffic. They are natural endpoints for moisture, liquids, crumbs, and other spills that provide molds, mildew, yeasts, and bacteria with a rich and practically continuous supply of nutrients. These microbes are in your home day in and day

Household Fast Facts

- Microbiologist Philip Tierno, PhD, has identified five places in the home that have the highest numbers of germs. They are (1) kitchen sponges and dish cloths, (2) the air blown from a running vacuum cleaner, (3) the washing machine, (4) a bathroom toilet during a flush, and (5) the kitchen trash can.
- Dr. Tierno's solutions to the problem spots listed above include: (1) changing sponges every one to two weeks or sanitizing with chlorine (bleach) and water; (2) replacing the vacuum bag once a month; (3) washing clothes with an antimicrobial laundry additive; (4) keeping the toilet lid down when flushing; (5) disinfecting the toothbrush regularly; and (6) disinfecting the inside of the trash can each time the liner is replaced.
- The cleanest-looking homes may be the dirtiest due to the spread of microbes with overworked sponges, grungy cleaning cloths, and dirty mops.
- The kitchen sink and drain are heavily contaminated with microbes. Faucet handles, refrigerator handles, and countertops are full of microbes, too, and usually have fecal bacteria.

out and are not a hazard unless they are given the opportunity. For instance, a barefoot walk with a cut on the bottom of your foot increases the chance for an infection. Infants may be more prone to infection as well because they spend a good deal of their day crawling on the floor. Floor travel increases a toddler's chances of breathing in or picking up a multitude of microbes.

Molds and mildew in carpeting release spores that cause allergic reactions, difficulty breathing, nasal and sinus congestion, eye irritation, and rashes. Asthma sufferers may be affected by just a small number of airborne molds. The size of a mold spore is smaller than a typical pollen grain, and it therefore may penetrate deeper into the lung passages than pollen.

Why are the numbers of people suffering from allergies and asthma on the rise? The molds found indoors are natural residents of the outdoors, but when they find favorable moisture conditions and nutrients, they can be very difficult to evict. In fact, the concentrations of airborne bacteria and molds are higher indoors than outdoors. Newer houses tend to be more airtight than older houses, so they don't receive cross-breezes and aeration. Air conditioning and heating systems recirculate indoor air, redistributing molds. A preference by many for climate-controlled homes with air conditioners, coolers, humidifiers, and vaporizers leads to generated moisture, and that helps mold grow. High humidity and little aeration is a mold-friendly combination. These are, unfortunately, conditions common in newer homes, schools, and office buildings. The EPA recommends that the relative humidity in homes range from 30 to 50 percent, and never exceed 60 percent.

Mold Growth and Control

Molds grow on almost any nonliving material if it is organic, that is, contains carbon. They can contaminate cellulose-based insulation, fireproofing, and ventilation system filters. Drywall in homes is often paper-lined, and this, too, is made of cellulose. Paperless drywall helps reduce, but does not eliminate, mold growth inside walls. In all cases, if water is available, the contamination will be worse. Porous (wallboard, fabrics) and semiporous (wood, concrete) materials provide crevices for growth and are difficult to decontaminate.

Day Care Centers and Nursing Homes

Day care or child care centers and elderly care facilities are similar because they house those at higher health risk than the normal population. The immune systems of the very young and the very old are not fully developed. The elderly may be weakened by disease or simple aging. Individuals are in close physical contact, and there are lapses in hygiene, so potential pathogens move easily from person to person. In day care, germ spread is further helped by sharing toys and putting toys and hands into the mouth. In addition, toddlers spend a lot of time on the floor. In both day care and nursing homes, fecal microbes contaminate meals as they might in any other setting. For all these reasons, infection is a major concern in day care and nursing facilities.

Health risks are compounded by an increase in antibiotic-resistant bacteria. Both sites house individuals that are in the same building or the same rooms for long periods of time. Some antibiotic-resistant microbes establish a population unique to the center. This unique setting and its conditions is a **microenvironment**. Constant exposure of children or the elderly to a microenvironment harboring resistant bacteria increases the chances for the spread of resistance among everyone, including staff.

American work schedules put increasing demands on the child care industry and its employees, and the trends are expected to continue. In addition, the Census Bureau predicts a dramatic shift toward an over-fifty-five population between now and 2025. These factors will only increase the importance of specialized caring skills and management.

In **day care**, enteric (diarrhea) and respiratory illnesses are the biggest problems, followed by ear infections and conjunctivitis. The incidence of diarrhea cases increases three and a half times when the center accepts children less than two years of age, on-site diapering being the main factor. Incidence increases threefold when meals are served at the center. Hepatitis A virus and rotavirus are the two main causes of diarrhea in day care children. *E. coli*, *Salmonella*, and *Campylobacter* contribute to diarrheal outbreaks to a lesser extent.

Ear infections are also a challenge, particularly since the causative *Streptococcus pneumoniae* (nicknamed pneumococcus) is becoming increasingly resistant to antibiotics.

Although young ones and toddlers are good at spreading germs on hands and toys, day care workers have also been shown to be a major vehicle for disease transmission. Children's Hospital Boston recently highlighted a distressing lack of knowledge by parents and staff members regarding the principles of germ transmission. Fewer than half the parents understood the connection between food preparation by an ill employee and outbreaks of diarrhea among children. The concepts of hand hygiene and person-to-person germ transmission were also not understood by about a quarter of parents and day care staff.

In **nursing homes**, the highest risks are from acute respiratory, urinary tract, and skin infections. *Streptococcus pneumoniae* is as prevalent here as it is in day care, but in the elderly it causes pneumonia. The risk of new infection is complicated by existing illnesses: diabetes, asthma, chronic illnesses, or cognitive disorders such as Alzheimer's disease. Patients with catheters, feeding tubes, or other invasive devices are also open to infection. In the normal aging process, the skin and mucous membranes begin losing integrity, and the opportunity for infection increases as these protective barriers become weakened.

As in day care, nursing home staff members contribute to the spread of a number of diseases. Their jobs require frequent touching of the persons under their care, and they often move directly from one person to the next. If the staff and cooks forget to remain vigilant against germ transmission, infections can quickly travel through a facility. To be fair, professionals that take care of young children or the elderly cannot control all the germ-spreading activities that happen on their watch in a single day. What is needed is a comprehensive understanding of hygiene and infection by parents and day care and nursing home staff. Educating children on good hygiene is also valuable at as young an age as practical.

Special Attention to Hygiene in Day Care and Nursing Homes

- Proper hand washing by staff AND children/elderly
- Washing hands at the beginning of each shift, before meals, after taking children/elderly to the restroom, and after diapering
- Disinfecting diapering tables after each use
- Keeping younger children separate from older children
- Segregating children with diarrhea or the elderly who are sick with a contagious condition
- Sending home or not accepting sick children
- Maintaining adequate nutrition for the elderly
- Training all staff and caregivers on preventive measures and personal hygiene

Mold growth can be in plain sight or it can be hidden inside walls. Leaks or flood damage inside walls is most worrisome because the "toxic mold" covers a large area before it is discovered. The "black mold" *Stachybotrys* flourishes after storms that cause water damage to buildings. A careful inspection may reveal an entire inner wall covered with black mold. *Stachybotrys* is famous, but it's not the only culprit. At least fifty other molds participate in structure damage, and all of them in large doses cause respiratory and allergic hazards. Some of the prominent building contaminants are *Alternaria, Penicillium, Aspergillus, Chaetomium, Cladosporium,* and a group called basidiospores.

In homes severely contaminated with mold, cleanup should be left to a professional, certified remediation company. Diluted bleach or antifungal chemicals used to be a do-it-yourself remedy, but the Occupational Safety and Health Administration (OSHA) no longer recommends them. OSHA's position is that they rarely remove every single mold spore, and large volumes of strong chemical cleaners only add to breathing hazards indoors. The professionals employ specialized wet-vacuuming and filtration procedures, and they use chemicals in a controlled and safe manner.

There are a few defensive measures one can take to prevent mold from growing worse. The areas near air conditioning units and furnaces should be checked periodically for stagnant water. Remove the water, dry the surface, and maintain a regular schedule for replacing filters and cleaning ducts. Repaint cleaned surfaces with mildew-resistant paints. Washing machines and dryers release moisture with each cycle. They should be vented to the outdoors, especially appliances that are in the basement. Cracks in the basement that are prone to leaks should be sealed, and other flood-reducing measures are always recommended for basements and crawl spaces. Fabrics and rugs should be removed regularly and shook and aerated outside. Regular rug cleaning by a professional service may also combat the mold problem.

Many brands of flooring and acrylic and nylon fibers in carpets are treated with antimicrobial compounds that give a modest inhibitory effect against molds and bacteria. There is debate among scientists regarding the benefits and hazards of using treated products

Antimicrobial Terminology

Biocide—Any chemical that kills living things, although it is often used to refer only to products that kill microbes.

Antimicrobial—Referring to a type of product (as an adjective) or a product (noun), one that kills microbes (or greatly reduces their numbers), including bacteria, yeast, fungi, algae, and protozoa. Most antimicrobial activity is that which destroys microbes on inanimate surfaces or in liquids. The following are types of antimicrobial products: antibacterial, antifungal, virucidal, bactericidal, and bacteriostatic. Germicidal means the same as antimicrobial.

Antibacterial—Capable of killing bacteria or reducing their numbers to "safe levels" as defined by EPA.

Antifungal—A type of product that kills fungi, including yeast and mildew.

Virucidal—Able to kill or significantly reduce numbers of viruses.

Bactericidal—A type of product that kills only bacteria.

Bacteriostatic—A type of product that *inhibits* the growth of bacteria, but may not kill them.

Antiseptic—A type of compound or product that kills or reduces the numbers of microorganisms on skin or mucous membranes.

Antibiotic—An antimicrobial compound, usually made naturally by bacteria or fungi, to kill other microbes.

such as these in the home. Many scientists and nonscientists argue that vigorous defense against all microbes doesn't make sense considering pathogens make up a very small percentage of the total. Each homeowner weighs the pros and cons of antimicrobial-treated building materials. Regardless of your choice, it is important to remember that household microbes are at your fingertips, in the air, and underfoot.

An antimicrobial chemical that seems to be everywhere in the home is triclosan (short for 2,4,4'-trichloro-2'-hydroxydiphenyl ether), a synthetic compound made of two benzene rings with an oxygen and a few chlorine molecules attached. It is also known as Microban, Biofresh, Irgasan DP-300, Lexol 300, Ster-Zac, and Cloxifenolum. Triclosan is in rugs and flooring to inhibit microbial growth, but it is a coating or an ingredient in almost any type of product that could possibly touch the skin, from plastic cutting boards to children's toys. Triclosan shows up in a surprising array of personal care products, too. Although it is known to irritate skin, eyes, and the respiratory tract, serious health problems from triclosan have never been proven in the thirty years it has been on the market.

Do antimicrobial products work? We investigate this further in the next chapter. For now, the answer is yes, they work. Some are strong chemicals that are highly efficient in killing germs; some are weaker. For an antimicrobial chemical to work, a minimum contact period is required. Contact time is the number of seconds or minutes a chemical must be in contact with a microbe in order to kill that microbe. Microbes are not killed instantaneously. If a product claims it kills germs "instantly," don't believe it. Triclosan's contact time is unlimited because it is formulated directly into a toy, bar soap, or mouse pad.

On the Job
Microbiologist Charles Gerba once observed that "nobody ever cleans a desktop until they start sticking to it." The average office has

21,000 microbes per square inch of surface—400 times dirtier than a toilet seat. Telephones can have over 25,000 microbes per square inch. Other dirty places are the computer keyboard (3300/sq. inch) and mouse (1700/sq. inch).

Coffee mugs are the Superfund sites of the workplace. Poorly rinsed mugs hold up to 300,000 bacteria. Hundreds to thousands have been recovered from "clean" mugs. Wiping out the inside of clean mugs with a moist sponge or cloth of questionable cleanliness only increases the bacteria. Pouring in hot coffee or tea does not eliminate microbes. The temperature of coffee, tea, soup, or other hot beverages is not high enough to kill them. Most are able to withstand much higher temperatures for several minutes.

Additional trouble spots in the workplace are microwave handles (10,000/sq. inch), water fountain handles (15,000/sq. inch), elevator buttons, and the buttons and surfaces of photocopiers and ATMs. The predominant bacteria found in these places are from the mouth. Fecal organisms also show up frequently at the office.

Unless these contaminants of the office get into food or into your mouth, they will remain harmless. The best way to reduce the spread of infection among coworkers is to stay home when you're sick.

Public Places

History tells us that as populations congregate, transmission increases. Buses, subways, elevators, and so on help infections spread.

Infectious microbes are also dependent on water. Moist or wet surfaces keep them alive longer. The sweaty palm against the bus or subway handrail delivers enough moisture to help bacteria survive there, and fecal bacteria have been recovered from handrails. One touch of your hand to a contaminated surface, and then another touch to the mucous membrane of the eyes, or to the nose or lips, is sufficient to infect yourself.

In your car, spilled food and an irregular cleaning schedule means an opportunity for microbes to grow. Cleaning, disinfecting, and vacuuming are as important in vehicles as they are in the home. As usual, this is especially true if a passenger has the flu or a cold. Car air conditioning systems may also harbor molds that are blown straight into faces. Studies have shown the following to be present in many car air conditioning units: *Penicillium, Cladosporium, Aureobasidium, Aspergillus, Alternaria,* and *Acremonium.* These are all common household molds and each can cause severe responses in allergy sufferers.

If you were to drop a cookie on a desk or the floor and pick it up within five seconds, how safe is it to eat that cookie? The answer is influenced in part by the amount of moisture on the surface. Therefore, dropping the cookie onto a table in the back room of a little-used wing of the library if you drop it onto the floor near a water fountain in the bus station.

As you venture farther from home and work, you may begin to notice other places where you have a better-than-even chance of picking up a microbe or two. Some of the locations you should view with caution are farms, animal barns, petting zoos, pet shops, greenhouses, shoe stores, airport screening areas where shoes are removed, laundromats, cosmetic counters with shared product samples, hair and nail salons, and barber shops.

Money
The word money is used with terms such as "ills," "dirty," and "filthy." Money has a bad reputation. Is money one of the dirtiest, most microbe-laden items we handle each day? Well, money is like any other object handled repeatedly by scores of people. Germs can travel on it.

There are few comprehensive studies on the germs found on paper currency. In order to recover microorganisms from a surface

Helpful Hints for Public Places

- In restrooms, pick the first or the last stall. The middle stalls get more traffic and therefore more germs.
- Public restrooms and Port-O-Potties get frequent washing with disinfectants and may be cleaner than you suspect. But the high numbers of people using them cause surfaces to become recontaminated quickly.
- Hotel room TV remote controls are not cleaned, they are handled numerous times by many people, and are often used after the bathroom and with moist hands. Carrying a small can of disinfectant or a package of alcohol wipes when traveling is not a bad idea.
- Handshaking transfers viruses and bacteria. Do not shake hands with someone who is obviously sick. If you must, wash your hands immediately afterward. The best advice is to avoid handshaking altogether (nearly impossible to do, I know). If you must shake hands with several people, and then immediately attend a meeting, avoid any contact with your face until you can wash. If doughnuts are offered, do not handle them with your bare hands. Remember, too, that you will transfer some of the viruses and bacteria to your pen, the tabletop, or computer.

such as paper, the best way of getting an accurate number of microbes involves destroying the sample. Don't expect to see data on one-hundred-dollar bills. Microbiologists have inspected one-dollar bills that had made their way through taxicabs and restaurants. Most bills hold a variety of bacteria and sometimes viruses. *Staphylococcus aureus, Streptococcus,* and Gram-negative bacteria such as *Salmonella* and *E. coli* have been recovered from bills.

Airplanes and Germs

Rumors of germs in airplanes circulate like the air inside a pressurized cabin. Airline travel has about the same level of microbial risk as other activities that involve crowding into a small space for several hours. Sharing close quarters with someone who's sick will increase your likelihood of becoming infected. Two points to remember about air travel are: (1) the plane cabin does not "fill up" with germs during flight, and (2) shared surfaces such as seats, armrests, overhead bin latches, and tray tables are good places for hands to pick up germs, which is no different from locations outside the airplane.

The Good News
- Airplane cabin air is exchanged about twenty times per hour. The air then flows through HEPA (high-efficiency particulate air) filters that remove 99.99 percent of particles between 0.1 and 0.3 micrometers and should be sufficient to remove almost all bacteria, fungi, and viruses larger than 0.3 micrometers. Even the smallish TB organism *M. tuberculosis* is about 0.5–1 micrometer in diameter and will be blocked by HEPA filters.
- Transmission of measles, SARS, colds, or flu during flight is thought to be extremely rare.
- Airflow on most modern planes enters above each row and moves downward, exiting beneath the seats. This minimizes the spread of germs beyond a few rows.
- Walking in the aisle causes brief disturbances to airflow and puts small dust particles into the air, but these particles return to normal levels within about three minutes.
- Airplane air is mixed with air from the outside. The ratio of recirculated to outside air can be up to 50:50. At 30,000 feet, the outside air is free of microbes.
- Although the airplane's bathroom faucet is covered with microbes, the toilet seat usually has very few!

The Bad News
- Airplane and air terminal bathrooms are high-traffic areas where microbes are on almost every surface. Though airport bathrooms are regularly disinfected, the high usage may override the benefits of cleaning.
- Airborne droplets will follow cabin airflow until exiting through

outflow ducts. Therefore, droplet transmission may be a concern, especially when seated next to a person who is sick or a carrier.
- Flights of eight hours or more increase the chances of getting infection from recirculated air.
- There has been at least one confirmed case and additional suspected cases in which disease was transmitted to healthy passengers during a flight.
- TB was transmitted to six healthy people during a 1994 Chicago-to-Honolulu flight. The method of transmission was found to be airborne droplets (sneezing, coughing) affecting people sitting near each other. Recirculated air was never confirmed as a factor.
- Airline meals, not air, spread foodborne infections if they are handled or cooked improperly.

Tips for Air Travel

Stores that sell travel accessories offer inexpensive products that may add a physical barrier against microbes. These include personal pillows and blankets, masks, seat wraps, and booties. Of these, a personal blanket may be best, especially if traveling with small children who will put airline blankets to their face. Face masks are not usually recommended by health agencies for air travel because only certain masks work against things the size of a microbe. The World Health Organization (WHO) has suggested that masks be considered if traveling to areas with a SARS outbreak. Infected persons showing SARS symptoms can infect others, and government agencies have published procedures for airlines to follow when transporting SARS patients.

In airplanes and terminals, travelers and crews are to follow the basic tenets of good personal hygiene, especially hand washing. Wash hands before eating and after using the restroom. Do not touch hands to the eyes, nose, or mouth. Do not share food or utensils. Finally, do your best to stay away from those who are sick with a cold or flu, quite a challenge in an airplane.

Many people are convinced that airplanes make them sick. Infection will commonly spread from a sick person to others when confined in a small area with many people touching the same surfaces. Consider, too, that travelers often experience stress. Vacations, business meetings, family visits, and holidays can be difficult occurrences. If you feel stressed during these times, imagine what your immune system is feeling.

Like paper currency, coins are handled hundreds of times in a single day. Bacteria and viruses have plenty of opportunity to ride along. *E. coli* and *Salmonella* have been recovered from U.S. coins, but the numbers vary widely. There is little scholarly opinion on whether disease is carried on money, but there is almost universal agreement that people who handle it and prepare food should wash their hands in between those two activities. Touching the ice dropped into cold drinks follows the same rule: wash before handling ice. A thorough wash with soap is needed, not just a quick rinse with water.

Handwashing

Wash with ample soap and warm water for twenty to thirty seconds, no longer. Wash all surfaces: the backs of hands, between fingers, fingertips, and up to the wrist. Rinse thoroughly and dry hands with a single-use towel. Turn off the water using a paper towel and immediately discard the towel.

Warm water is more effective than cold for dissolving the surface-active ingredients in soap, which help loosen and wash away dirt. Germs are on dirt as well as your skin. Washing should be thorough. Rinsing and repeating is a good idea. Hot water is not recommended because it removes the oils that protect the skin. Washing several times a day is recommended for removing the constant onslaught of germs, but it also has the shortcoming of causing dryness. Hands that are dry to the point of cracking are vulnerable to infection. Use lotions or moisturizers for dry or cracked skin.

U.S. coins are made of copper, nickel, and zinc. These metals are known to inhibit the growth of bacteria and molds—mariners for centuries copper-plated the bottoms of their ships to slow the attachment and growth of algae—but the metal works best when it's bound to some other compound. There is no solid evidence that metal coins repel germs, but we know a coin is not a hospitable place for a microbe to grow. Both copper and zinc have been used for years in products to kill fungi. Compounds containing copper are used in wood, fabric, and paint preservatives and in fungicidal disinfectants. Zinc oxide is used by the paint industry as a preservative.

Paper currency is unlike coins because it has a porous surface made of fibers and inks that bacteria can digest. Like all other porous surfaces, the paper offers thousands of hiding places in which germs ride along. The U.S. Treasury Department has experimented with antimicrobial additions to money for preserving the bills and to reduce the potential hazards spread by them. At this time, however, there are no antimicrobial chemicals in currency.

There seems to be no place at home or at the workplace where you can be free of microbes. They are everywhere. Yet for all the concerns about germs on money, in the bathroom, and around the kitchen sink, you go through the vast majority of your days and years healthy and infection-free. You can attribute that to two things. First, the number of harmless microbes on earth is far greater than the amount of dangerous ones. Second, you have now learned the importance of good personal hygiene. At its most basic, wash hands frequently and thoroughly, and do not touch your unwashed hands to your face.

The Microbes on the Surface of Your Body

Humans and other mammals are built like tubes. The outer surface of the tube is made of epidermis; the inner surface is the lining of the digestive tract. It is normal to have a population of bacteria, yeasts, and molds on your skin. It's better to have these microorganisms on you than to not have any at all. The bacteria normally found on the skin are one of the body's first lines of defense against pathogens. These native bacteria outcompete pathogenic bacteria and yeasts for nutrients. Some excrete antimicrobial compounds that further discourage pathogens from settling in. They establish communities on specific areas of the body, influencing the microconditions found in these places: acid content, protein amounts, and waste products. For example, the group of skin bacteria called propionibacteria produce fatty acids, which inhibit many other Gram-positive species. Your native flora, therefore, have a type of squatters' rights on your body.

Skin

The skin is exposed to the air, so it is no surprise that many of the microbes that cover your body are aerobes; they use oxygen to stay alive. More surprising is that the predominant bacteria on the skin are anaerobic, about 100 anaerobes to one aerobe. Anaerobes thrive on the epidermis because it has abundant small places known as microenvironments that are depleted of oxygen. For instance, the depths of a pore represent an oxygen-scarce microenvironment. This is one of several microenvironments on the body in which specialized bacteria take advantage of a very narrow range of conditions for their growth.

Normal flora vary from person to person. Factors that influence an individual's normal flora are owning pets, handling horses and livestock, frequent swimming in chlorinated water, antibiotic treatment, diseases and illness, burns, skin infections, use of antimicrobial soaps or antiseptics, hospitalization, and different genetic and familial factors. Even considering these many variations, most people have in common a basic set of native flora.

The epidermis offers an assortment of living conditions for

microbes. The arms, chest, back, and legs are often referred to as dry and sparse deserts. Bacteria such as *Staphylococcus epidermidis* and *Propionibacterium* live there. The nostrils, underarms, groin and genitals, and parts of the feet are the rainforests of the human body. These areas have the two bacteria found in dry places plus large numbers of *Staphylococcus aureus* (nostrils) and *Corynebacterium* species (underarms). The groin and urogenital regions are home to those four plus *Lactobacillus, Streptococcus,* a group known as diptheroids, and an additional variety of Gram-negative bacteria. The adult female urogenital region also harbors *Candida albicans* yeast.

Showering and washing rinses off millions to billions of skin microbes. The microbes attached to dead skin cells and dirt particles are the easiest to wash away. But many stay behind after the shower ends. They hide in skin's microscopic crevices. Many also have mechanisms for remaining firmly attached to surface skin. After washing, it takes only a few hours for the population to begin replicating. The entire community returns to its original numbers after about twelve hours.

If camping or traveling or doing anything else that precludes daily showering, do the microbes on your skin grow to incredibly high numbers? You may already have guessed that something prevents this from happening. Bacteria and fungi in nature grow to the capacity of their environment. On the skin, bacteria multiply until the nutrients needed for their growth are depleted. Growth stops also when the bacterial numbers become high enough to force cells to be close to one another. Certain bacteria ward off the others near them by excreting compounds that inhibit their neighbors from growing any further. In this way, your native skin flora help protect against a team of pathogens establishing a beachhead on your body.

Lactobacillus is an example of a beneficial species. *Lactobacillus* is in the urogenital region and lowers vaginal pH to 4.5 or less, conditions where it thrives but invading pathogens struggle. The pH of an environment is a measure of acidity. Acidic conditions are of pH less than 7, basic conditions are greater than 7, and 7 is neutral. When certain antibiotics are taken (rifampicin, tetracycline, chloramphenicol, penicillin, ampicillin, cephalothin), *Lactobacillus* species disappear, the pH rises, and *Candida* yeast that are normally present, but in low numbers, multiply. Thus, a harmless skin inhabitant becomes a pathogen as conditions are altered to its advantage.

Body odor from the underarms is not a disease, but we treat it as such by attacking it with a veritable chemical arsenal of soaps, deodorants, and antiperspirants. The offending odors come from the normal activities of *S. epidermidis* and groups known as the corynebacteria and diptheroids. *S. epidermidis* is rare among the staphylococci for its ability to convert underarm sweat into an annoying odor. As the underarm bacteria digest the sulfur-containing amino acids in sweat proteins, new sulfur compounds are released. Many of these compounds are volatile (evaporate quickly into the air as a vapor), and are therefore easily detected by the nose.

The bacteria of the underarm are particular about what they eat. They produce odor only from one of two types of sweat the human body produces. Eccrine sweat glands are found all over and produce a watery liquid for cooling the skin surface. Apocrine glands are only in a few locations, including the underarms. They exude a thicker sweat loaded with peptides (short strands of amino acids) and salts. Underarm bacteria feast on these ingredients.

Body odor from places other than the armpits builds up during the period between showers. This happens even when sweat glands aren't secreting. Microbes break down the oils and triglycerides in sebum to fatty acids. As other microbes further digest these compounds, their volatile fatty acid by-products disperse into the air and spread their sour odors.

Athlete's foot (or tinea pedis) fungus is caused by a filamentous fungus (when it grows it spreads by forming strings, or filaments, made of long cells joined together), *Trichophyton mentagrophytes.* Like many fungal infections, it lives on dead skin cells that have completed their normal process of migrating outward from the deep epidermal layers. If the skin becomes irritated from scratching, *Trichophyton* will invade deeper into the skin layers. Because fungus cells are more related to mammalian cells than are bacteria, the drugs that kill fungi often have devastating effects on human cells as well. This makes fungal infections of the skin very difficult to cure.

Eyes

The eye is another place where yeast and fungi are dangerous. Normally the conjunctiva and cornea are impermeable to invasion by microbes. But even a minute injury to the eye leaves it vulnerable to infection from bacteria, viruses, or fungi.

Contact lenses do not provide protection. Soft contact lenses themselves are susceptible to contamination if not properly handled. If a contaminated lens is placed directly on the eye's surface, infection is almost guaranteed. Some of the yeasts that contaminate lenses are: *Candida, Rhodotorula, Torulopsis,* and *Cryptococcus.* Additional problem fungi are *Aspergillus, Penicillium, Fusarium, Alternaria,* and *Cladosporium. Fusarium keratitis* has recently been linked with certain lens care products. The Food and Drug Administration (FDA) provides regular updates on the status of this infection plus other tips on contact lens care.

General contact lens safety tips include the following: remove immediately if the eyes are irritated or red; clean and disinfect lenses properly and according to the directions for your lens type; replace the storage case every three to six months; use fresh contact lens solution and never reuse; avoid cleaning lenses with nonsterile distilled water, tap water, or saliva; and wash hands properly before and after handling lenses.

Body Products

Product	Active Ingredients	How It Works
Personal Care Products:		
Toothpaste	Sodium fluoride	Rebuilds tooth enamel decayed by oral bacteria
Tartar control toothpaste	Sodium mono-fluorophosphate, triclosan	Interferes with nutrient uptake by bacteria; triclosan poisons bacterial membrane
Mouthwash	Cetylperidium chloride, alcohol, menthol, methyl salicylate	Chloride and alcohol disrupt membrane; salicylate removes excess skin cells; menthol for odor
Antiperspirant	Aluminum zirconium tetrachlorohydrex	Plugs sweat glands, and inhibits staphylococci
Deodorant	Poloxamine 1307, disodium EDTA, glycol compounds	Inhibits bacteria growth
Contact lens cleaner	Edetate disodium polyquaternary-1	Preservative and prevents bacteria and yeast from attaching to lens
Acne cream	Benzoyl peroxide	Disrupts fatty acid use by *Propionibacterium acnes*
Anti-dandruff shampoo	Selenium sulfide, coal tar, pyrithione zinc, salicylic acid, ketoconazole	Coal tar and salicyclate dislodge dead skin. Others inhibit *M. furfur*

Product	Active Ingredients	How It Works
Athlete's foot spray	Miconazole nitrate	Inhibits fungi
Foot powder	Menthol, corn starch, sodium bicarbonate	Starch and bicarbonate absorb moisture; menthol for odor
Medicated adhesive bandages	Polymixin B sulfate, bacitracin zinc	Destroy bacteria cell membrane and wall; bacitracin kills bacteria, polymixin kills bacteria and fungi
Antiviral tissues	Citric acid + sodium lauryl sulfate	Citric acid interferes with enzymes. SLS destroys proteins
Antimicrobial Q-tips	General antimicrobial treatment on cotton swabs	Used to protect integrity of the swab, not for killing bacteria in the ears

Antiseptics:

Benzalconium chloride	General antimicrobial activity
Isopropyl alcohol	Destroys microbial membranes and proteins
Hydrogen peroxide	Powerful oxidant poisons bacteria, yeasts, and viruses
Iodine + Alcohol	Destruction of microbial cell components

Table 3.1. Antimicrobial personal care products.

Hair

Scalp microbes are similar to normal skin flora with the exception of *Malassezia furfur*. This yeast is thought to be one of the causes of dandruff and seborrheic dermatitis. It's difficult to rid the scalp of *M. fufur* and many shampoo formulas have been invented over the years to keep it under control. Some experts believe coal tar is head and shoulders above the rest. Coal tar and salicylic acid may be most valuable as chemicals that help shed dead scalp cells. With the skin cells go an abundance of yeasts. But coal tar and other antidandruff compounds in shampoos are not aesthetically pleasing. Unfortunately dandruff is difficult to prevent or cure.

The bowling pin-shaped *M. furfur* is closely related (some microbiologists have suggested they are the same) to *M. canis,* the yeast that causes greasy, smelly skin on dogs and a dark waxy buildup in their ears.

The genus *Malassezia* also goes by the name *Pityrosporum*. Both names refer to the same microbe.

Microbiology of the Mouth

The human body is made of 10 trillion cells and includes bone, muscle, nerves, skin, connective tissue, and other specialized tissue. The number of bacteria on and in your body exceeds *100 trillion.* Your body is like planet earth; its bacteria outnumber every other living thing. In mammals, the mouth and the intestines house the greatest concentrations of microbes.

The mouth is unique because it's the only place on the body that provides a hard surface to which microbes attach. There are staggering numbers of bacteria in the mouth—10 million to 10 billion per cc of saliva or gram of dental scrapings. (A cc is a volume equaling one cubic centimeter—about the size of a sugar cube. A gram is the weight of one cc of water or roughly the weight of a one-dollar bill.) Hundreds of species have been isolated, and there are undoubtedly many that have not been recovered and identified

(Figure 3.1). Though the total amount of oral microbes is huge, the number is not as important as their types and their roles.

Six features of the mouth determine the type of microbial growth found there. The first is a constant warm temperature, 95 to 96.8 degrees F (35–36 C), that coincides with most bacteria's optimal range. The second is the reduced-oxygen conditions inside the mouth that favor some microbes and inhibit others, and greatly

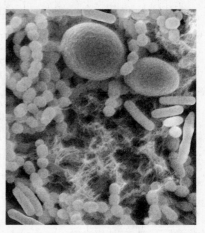

Figure 3.1. A few of the many oral microbes: *Streptococcus mutans* (strands of round bacteria) and the sticky material it makes during the first steps in plaque formation; *Bacteroides*, the rod-shaped bacteria; and *Candida albicans*, the large oval buds; magnification, x2000. Copyright Dennis Kunkel Microscopy, Inc.

influence the microbial activities taking place there. Nutrient supply in the oral cavity is another factor. Nutrients may come from the body itself, from saliva and from crevicular or gingival fluid, a fluid different from saliva that is secreted from the cells in the gingival crevices between the teeth. Diet also supplies a rich mixture of nitrogen compounds, carbohydrates, fats, water, and minerals. Fourth, the saliva provides conditions that are slightly on the acid side of neutral. When sugars are rapidly digested, the mouth becomes more acidic. This shifts the composition of the bacterial species for a period of time. Fifth, the mouth offers a variety of invaginations, crevices, and fissures that gives many species a chance to attach to oral surfaces and establish stable communities. Sixth, there is a relationship between oral bacteria and the body's immune system, allowing certain bacteria to survive there but nowhere else on the body.

Saliva has fewer bacteria than is found on the tooth surface or in gingival crevices. When oral bacteria attach themselves to a surface, they adapt to the mouth's periodic rinses, swallows, and splashes. They then form a stable conglomerate of over 400 species that collectively share jobs like nutrient extraction from food, nutrient storage, and building a protective covering. These surface-attached communities are called **biofilms**, and whether they are found inside or outside the body they are more resistant to destruction than free-floating bacteria. Biofilms containing *Streptococcus, Haemophilus,* and *Moraxella* bacteria inhabit the middle ear and are thought to contribute to chronic ear inflammations in the young. Other natural biofilms are those found in flowing streams and rivers, the inner surfaces of water distribution pipes, and inside toilet bowls!

Dental caries, or cavities, is a disease state of the mouth. *Streptococcus mutans* and other oral bacteria break down dietary sugars and carbohydrates and form weak acids. Food debris and saliva, bacteria and their acids all combine into a biofilm known as plaque, which starts to accumulate on teeth within twenty minutes after eating. If not removed, plaque holds the acids against the tooth surface, enabling the acid to corrode the enamel, thus beginning the decay process. Brushing, flossing, regular cleaning, and attention to diet are the best preventions for dental caries. Fluoride compounds in toothpaste remineralize the teeth by forming replacement enamel that is more resistant to decay than natural enamel.

Bad breath or halitosis is not usually a serious malady but many people treat it as such. It's caused by the microbes that live in oxygen-depleted conditions. Anaerobes are predominant in the mouth. Some anaerobic bacteria are actually **facultative anaerobes.** They are termed facultative because they use whatever tiny amounts of oxygen are present in their surroundings, and then convert to an anaerobic system once the oxygen has been used up. Without oxygen facultative anaerobes keep right on growing, and some do better without it than they do in fully aerated conditions.

During sleep, facultative anaerobes in the mouth deplete the oxygen in oral microenvironments such as the areas between teeth, the gum pockets, and invaginations on the tongue. Facultative bacteria plus the true anaerobes (known as **obligate anaerobes**) then live in oxygen-free conditions for the rest of the night. During anaerobic growth, they give off by-products that are smelly in very low concentrations. Therefore, the normal anaerobic bacteria in the mouth are the main cause of "morning breath."

Bad breath may also indicate unhealthy nonmicrobial conditions. An example is ketosis. This condition occurs during fasting or starvation or in some cases of diabetes mellitus. If carbohydrates are not available due to diet or a lack of insulin, the body's fat reserves begin to be liberated for energy. Fatty compounds travel in the bloodstream to the liver, where they are degraded to a trio of compounds known as ketone bodies. When these volatile ketones build up in blood and other fluids, they are given off in the breath. A characteristic fruity "ketone breath" is the result, different than microbial halitosis.

The back of the tongue and the gingival crevices are sites associated with oral malodor. Unlike saliva or the tooth surface, the tongue and gums offer pockets and invaginations for bacteria to escape removal by brushing or rinsing. The back of the tongue is also a great place to catch postnasal drip, another food source for bacteria. Plaque and tartar at the gum line contain anaerobic bacteria that add their own malodorous compounds to the mix.

Bad breath seems to be one of Americans' great fears. The money spent annually on toothpaste totals over $1.8 billion, with over $950 million on toothbrushes and floss, $740 million on mouthwashes, $715 million on oral-care gum, and $625 million for breath mints and fresheners. Noteworthy also is the recent establishment of ISBOR, the International Society for Breath and Odor Research!

Bad-smelling compounds are products of normal microbial

breakdown of food, especially proteins. Proteins are made of amino acids. Amino acids always contain nitrogen and often sulfur. Volatile nitrogen and sulfur-containing compounds are some of the smelliest things on earth. Believe it or not, there are microbiologists who study the oral bacteria that release the worst smelling compounds. Some of the compounds that have been identified are hydrogen sulfide (rotten-egg smell), cadaverine (decaying corpse smell), putrescine (decaying meat smell), isovaleric acid (odor of sweaty feet), and methyl mercaptan and skatole (feces-like odors). Little wonder so much money is spent on oral products.

There are ways to prevent bad breath: gentle cleaning of the back of the tongue, eating a good breakfast to stimulate saliva flow and cleansing, preventing dry mouth by drinking sufficient liquids or chewing gum, using mouthwash, and brushing and flossing after meals.

Microbiology of the Intestines

Human digestive tract contents make up about 8 percent of the body's weight. Within the entire tract, the mouth and intestines combined hold the largest concentration of the body's microbes. The overall concentrations of bacteria are high in the mouth, decrease in the stomach, and then increase again in the large intestine and rectum. Passing through the stomach is the most hazardous part of microbes' travel through the entire digestive tract. The stomach contains powerful gastric acid that kills almost all incoming bacteria. Numbers of bacteria can drop to only a few or a few dozen per cc of stomach juices. Those that manage to survive as they transit the stomach do so by hiding in pockets of undigested food particles that protect them from the acid.

Microbiologists study gastrointestinal bacteria by taking samples from the mouth and from stool. Reaching the stomach and small intestines for samples is understandably more difficult. That's why there are more published articles on human oral and fecal (or colon) microbes than on gastric and upper intestinal microbes.

The lucky survivors of the stomach enter the small intestine, where conditions are favorable for renewed growth, and their concentration starts to rise. Bacteria reach their highest numbers in the large intestine or colon, billions to trillions of cells per gram of fecal material.

Over four hundred species of intestinal bacteria provide humans with surprising benefits. In addition to (a) making vitamins, they (b) digest fibrous foods (vegetables and fruits), (c) contribute nitrogen in the form of amino acids the body uses to make proteins, (d) stimulate the immune response to nonnative microorganisms, and (e) compete with pathogens for attachment sites along the intestinal lining. The wonderful relationship between you and your gut bacteria is called **mutualism**. In a mutualistic association, both you and your microbes receive benefits.

Commensals and Parasites

When two organisms such as humans and their microbes live together, and one is benefited while the other is neither benefited nor harmed, the relationship is called **commensalism**. For example, Candida is a commensal microbe unless antibiotics or other conditions affect the bacteria-yeast balance. Then the Candida changes from being a harmless organism to a pathogenic one. In **parasitism**, the microbe benefits but the host is harmed. The fungus Trichophyton is a parasite when it causes athlete's foot.

When you eat a small amount of spoiled food or swallow a questionable morsel, you may not get sick. Why not? Aren't bad microbes as fond of your intestines as are the good microbes? Many of them may be, but the cells lining your digestive tract are smart in differentiating friends from foes. Your body's immune system doesn't kill your normal flora. Native microbes are immune to your immunity, you might say, because they produce chemicals that prevent the immune system from turning on. But pathogens such as foodborne bacteria don't have this ability. When pathogens arrive, your body recognizes them as invaders. A series of events are then activated to flush the pathogen or its toxin from the intestines or expel it from the stomach. Although the result is uncomfortable—retching, vomiting, diarrhea—your body is doing the right thing in ridding you of the interloper.

In the digestive tract and on the skin, the body's defenses work particularly well against small numbers of bad microbes. In a normal day, you are constantly exposed to bad apples (the microbial kind), but normal body flora and your immune system eliminate the threat without your being aware. How does this relate to the Five-Second Rule? Although a cookie will undoubtedly capture a few microbes from even a clean-looking floor, a healthy person's immune system will resist minor assaults from microbes whether they are harmless or dangerous.

The Daily Battle
Revisit a typical schedule of daily events. Throughout the day there are hundreds of activities that are based on microbiology. Some of the microbes you encounter are potentially dangerous and are controlled by preservatives or antimicrobial products. Despite the potential hazards listed here, most of the microbes you meet throughout the day mean you no harm.

6:00 AM *Alarm rings* Bedsheets potentially contaminated with fecal microbes.

6:05 AM *Brush teeth* Toothpaste and mouthwash reduce the number of oral bacteria and plaque-forming species.

6:10 AM *Shower* Soap and water remove dead skin cells, oily sebum, and some skin bacteria and yeasts; dandruff shampoo kills a percentage of scalp yeast.

6:20 AM *Bathroom* Antiperspirant/deodorant lowers the number of underarm odor–producing bacteria; foot powder containing menthol reduces moisture to inhibit fungi and reduces odors from bacterial and fungal growth.

6:22 AM *Bedroom* Clothes, carpet, and shoes contain bacteria and fungi. Many carpets are treated with antimicrobial additive. Shoe inserts include carbon granules and sodium bicarbonate to reduce bacterial and fungal odors.

6:35 AM *Kitchen* Milk is pasteurized to delay spoilage by bacteria; eggs may carry live *Salmonella* until cooked; ham is preserved with nitrites; bread for toast is the product of fermentations by *Saccharomyces* yeast. Foods that are refrigerated to prevent *Pseudomonas,* lactic acid bacteria, and others from causing rancidity and souring of milk, cream, and butter. Refrigeration also delays bread mold growth. Daily vitamin contains B12 from *Pseudomonas,* vitamin C from *Acetobacter,* and riboflavin from *Ashbya gossypii* fungus.

6:55 AM *Utility room* Change litter box to eliminate odors from *Proteus* bacteria breakdown of urine. Put out garbage to remove spoilage organisms and their odors.

7:10 AM *Bus stop* Other commuters, possibly with an infectious bug.

7:15 AM *Bus* Handrail, seats, and coins contaminated with bacteria and viruses. Use hand sanitizer to wash away potential infectious microbes.

7:45 AM *Office* Coffee mug, desk, and computer contaminated with bacteria, fungi, and possibly viruses.

8:30 AM *Restroom* Toilet, toilet stall, aerosols, sink, and faucet handles hold microorganisms. Wash hands before returning to desk.

10:00 AM *Snack* Yogurt made from pasteurized cow's milk that is inoculated with *Streptococcus and Lactobacillus* bacteria.

12:00 PM *Lunch* Hamburger potential source of *E. coli* and other fecal bacteria. Proper cooking kills them. Salad potentially carries fecal contaminants. The ice cubes in the iced tea potentially add *Pseudomonas* and coliforms. All food a potential source of illness if cooks/servers do not follow good hygiene and food handling.

12:50 pm *Soda machine* Large amounts of bacteria at the liquid dispenser.

1:00 PM *Conference room* Cream-filled doughnut potentially harboring *Lactobacillus,* if not properly refrigerated before serving.

3:00 PM *Office* Phone covered with bacteria and possibly viruses.

5:15 PM *Gym* Bacteria and virus transfer by way of equipment; pool water uses chlorine to kill contaminants; shower floors carry bacteria and athlete's foot fungi.

6:25 PM *Train* Money may carry small amounts of microbes.

6:45 pm *Gallery opening* Air conditioning disperses several mold species and may harbor *Legionella* bacteria. Bacteria and fungi on paintings eat the pigments, pallet components, and fungi infiltrate wood frames.

7:30 PM *Home* Bacon, lettuce, and tomato sandwich may have fecal contaminants in the lettuce, but bacon is preserved with nitrate compounds. Pasta salad kept refrigerated until eaten to avoid overgrowth of *Salmonella* and spoilage bacteria. Sponge used to clean up spreads germs around.

Bathroom tile grows *Serratia* bacteria and *Aspergillus* mildew. TV remote potentially holds fecal contaminants. The cat is a potential source of *Toxoplasma* protozoa.

9:00 PM *Laundry* Washing machine spreads fecal bacteria throughout the load. Bleach and antimicrobial laundry detergents help reduce contaminants.

10:15 PM *Brush teeth* Floss removes food that enhances the formation of plaque by various streptococci.

10:25 PM *Bedroom* Antiviral tissue kills cold, flu, and other viruses.

Summary

There are few activities in a twenty-four-hour period that do not have a connection to microbiology. The harmful or merely annoying microbes receive the most attention, a type of "squeaky wheel" approach to microbiology. But the good microbes far outnumber the dangerous ones, and dangerous microbes don't automatically start infection. You have three excellent defenses at your disposal to keep you safe from small-scale assaults by pathogens: your native flora, your immune system, and good personal hygiene. To become dangerous, a microbe needs a perfect storm of virulence factors, host susceptibility, and opportunity. The outcome can be devastating when these three elements unite.

4

Eat, Drink, and Be Microbial

We now arrive at a very delicate point in these researches. I would like to speak of the relationship which exists between the sugar and the yeast.

—Louis Pasteur

Your grocery bag contains the highest amount of microbes you bring into your home each day. With each swallow of water you ingest additional microbes, mostly bacteria.

Microorganisms in foods and drinking water have been with us since the dawn of humanity. Early societies confronted the challenge of ensuring a pathogen-free diet—long before the invention of the microscope—by heating, drying, smoking, salting, pickling, or fermenting foods and beverages. Salting meats has been practiced since the fifth century B.C. Preservation methods used by our ancestors have been improved, perhaps by scientific study but more likely through years of trial and error. Nevertheless, traditional techniques in preservation remain in use today with surprisingly few modifications from those practiced by early civilizations.

Scientists gained details on microbes in water with the introduction of microscopes, but beginning in 50 B.C. the Romans recognized the health benefits of clear, running water. Roman metropolises were served by aqueducts that supplied freshwater for bathing and drinking. Sewer systems washed away used water. And in the baths, a sprinkling of spices and fragrances may have served as a rudimentary treatment method. The Romans didn't know the finer points of antimicrobial action, but they may have intuitively recognized the benefit of adding spice oils and various flower and plant extracts to bath water.

Unfortunately, the advances in water quality pioneered by the Romans went largely ignored in following centuries. There are ample and unsavory examples of uncontained sewage gurgling through streets in the Middle Ages. It wasn't until 1854 that public health officials began using chlorine compounds for treating sewage and potable (drinkable) water. Treatment with chlorine and hypochlorites (compounds that combine a chlorine molecule with hydrogen and oxygen) has not changed significantly over centuries.

"From Toilet to Tap"
The Water Cycle
Water circulates on earth as blood flows through the body, in a cycle. It evaporates from oceans and lakes, condenses among the clouds, falls as rain into lakes or to the ground. It evaporates back into the atmosphere or tumbles toward the ocean. A single drop of rainwater sometimes travels thousands of miles from lakes to rivers, and on to bays and estuaries leading to the sea. Wildlife and humans are but a brief detour in the cycle.

The water we drink either moves through the digestive tract and exits in feces or it is absorbed through the lining of the small and large intestines. The small intestine alone absorbs up to two and a half gallons each day but has the capacity to absorb much more. Drinking water and beverages account for about two-thirds of absorbed water. Water diffuses through the intestinal lining into the blood, which distributes it to tissues to maintain cellular integrity and all cellular functions. Many nutritionists consider water to be a dietary nutrient just the same as protein, carbohydrate, fat, minerals, and vitamins. Eventually water leaves the body by way of the kidneys (about 60 percent of the total daily intake goes out in urine), the feces (about 8 percent), and sweat (about 4 percent on average). The remaining 30 percent of absorbed water leaves the body by evaporating from the lungs or diffusing through the skin. Thus, your body's water returns to the atmosphere or begins a journey to a sewage treatment plant.

Sewage Treatment

As you might expect, sewer water holds large amounts of microorganisms shed from humans and other animals. It also contains organisms found in nature that are washed into streams during rains. Sewage is the resulting mix of wastewaters, rain and natural runoff waters, irrigation waters, and liquid and solid wastes from houses, towns, and various sources in the wild.

Sewage is cleaned at all treatment plants following standard steps. First, wastewater enters the treatment plant and flows into massive open-air tanks where, second, fine particles are chemically coagulated to form larger particles that settle to the bottom. Some small particles remain suspended in the partially clarified water. Bacteria, viruses, cysts, and other organic (containing carbon) materials remain attached to the particles. The suspension then flows to aeration tanks, where it mixes with good bacteria whose sole job is to eat up as much organic material from the small particles as possible, leaving behind a heavy sludge. The paradox of sewage treatment is that as microorganisms are removed, billions of other helpful bacteria are added for digesting wastes. The clear water is mixed with chlorine compounds, which kill all remaining bacteria. Meanwhile, the sludge from the aeration tanks is pumped to an enclosed tank where anaerobic bacteria (those that live without oxygen) complete the breakdown, giving off methane gas as an end product. At night you may notice a flame at the top of a tall tank at the sewage plant. That's methane being burned away. By continually getting rid of the methane, reactions inside the anaerobic tank can continue. If too much methane were to build up, the digestive reactions would stop.

Note also that sewage plants are located at or close to the lowest sea level in town, making it easy to gather runoff and urban wastes.

Sewage treatment removes fecal matter containing bacteria and viruses. There are additional living things in water: algae, protozoa, fungi, worms, insects, and mollusks. Two protozoa shed in cattle, sheep, and wildlife feces are also found in the water heading into a treatment facility. They are *Cryptosporidium* and *Giardia,* the bane of campers who drink directly from woodland streams.

Chemical disinfection takes care of viruses, bacteria, and algae. Toxins and other dissolved or suspended materials are captured when the water flows through carbon powders and sophisticated filters. Filtration (passing dirty water through tiny pores so that only clean water emerges and large particles are captured on the filter) removes the big guys, that is, worms, insects, and so on. But Crypto and *Giardia* form small robust cysts; Crypto cysts (sometimes called oocysts) are 3–6 micrometers in diameter, *Giardia's* are 8–16 micrometers. These are big compared with bacteria, but still microscopic. Crypto cysts occasionally find their way past the settling-filtration process, especially when heavy rainfall exceeds the sewage plant's capacity, and, because it's resistant to chlorine, Crypto can emerge from treatment plants undamaged.

The technology of sewage treatment has advanced to the point where the effluent from the most sophisticated treatment plants is much cleaner than almost any natural water source. Treated water is usually returned to the ocean, bays, or other large tributaries, or is used for irrigation. In California, San Diego's water district is resurrecting a campaign that was roundly criticized some years ago. Taking pride in their technical expertise for recovering scarce water and cleaning it for human use, officials came up with the catchy slogan, "From Toilet to Tap." The citizens of the district did not, however, appreciate the scientific marvels involved and pictured only a filthy type of water about to pour from their faucets. The Toilet to

Tap campaign was derailed after the hearty public outcry. Recently, the city has reinitiated the proposal for recycling treated wastewater. It is to be pumped to reservoirs to mix with natural freshwater before being distributed to homes and businesses. The politicians have wisely renamed the plan the "Water Reuse Study."

Drinking Water Treatment

A glass of cool, clear water connotes restorative rewards and health. Hold onto that thought. Water also delivers to you a daily stream of bacteria.

Drinking water treatment is similar to wastewater treatment: a series of settlings, filtrations, and disinfections. Water for drinking and bathing is disinfected twice, once at the beginning of the process and again at the end before entering the pipes that lead to homes and offices. Treatment plants receive water from reservoirs, rivers, lakes, or canals. After treatment, the water is stored in tanks, usually at the highest points in town above sea level.

Chlorine gas and chlorine hypochlorite are commonly used for water disinfection; chloramine, chlorine dioxide, and chlorine–chlorine dioxide mixtures are gaining in popularity. Some towns use ozone. The monthly report from your water district describes the method used in your town. Water-quality scientists have determined that chlorine and hypochlorites put small amounts of pollutants into the environment, and so they continue to devise new ways for disinfecting water *and* protecting the environment.

The closer you live to the treatment plant, the fewer bacteria are likely to pour from your tap. Even so, microbial concentrations tend to fluctuate from house to house and from hour to hour within a house.

Two main reasons for the fluctuation are the quality of the distribution system and the presence of biofilm. Corroded pipes and those with pockets of stagnant, nonflowing water hold high

Water Disinfection

Drinking-water treatment plants use different methods to disinfect water, alone or in combinations:

Sodium hypochlorite (bleach) Effective but loses strength within a few hours

Chlorine gas Effective but corrosive and potentially explosive

Chloramines Work slower than chlorine hypochlorite but last longer

Ozone Gas made of three oxygen molecules; effective and there is no chlorine taste and odor

Irradiation Ultraviolet (UV) light kills microbes in water but doesn't work well on cloudy water

Silver Popular in Europe; less effective than chlorine

amounts of bacteria. New buildings and homes and new restaurants often have stagnant water in their piping. If you're using the tap or fountain in a recently constructed building, run the water for at least two minutes. The extra time isn't necessary once the house or office starts to support normal daily activities. As you'll soon see, bottled water may be no safer a choice than tap water. In new restaurants— well, there isn't much you can do except order a soft drink or a glass of wine!

Biofilm is a complex mixture of bacteria—sometimes it also contains fungi and algae—that adheres to the wall of any conduit with flowing liquid (Figure 4.1). It grows inside water distribution lines and also on the hulls of boats and on the rocks and pebbles in streams. Water microbes have adapted to life with few nutrients. Biofilm bacteria are even more specialized because they have adapted to living in the liquids flowing past them, sometimes flowing at a great speed. To avoid being washed away with the flow, they produce a sticky mass of large carbohydrate-like compounds that holds them in place on a surface. This film also serves the microbes embedded in it as storage for nutrients drawn from the waters rushing past. In addition, the glob protects biofilm microbes from disinfectants. Once attached, the mixture is extremely hard to remove from any surface it has claimed. Chlorine barely penetrates biofilm, so it doesn't kill most of the well-protected microbes inside. In addition, chunks of biofilm often break away from pipes and flow downstream, causing tap water to contain few bacteria one minute and hundreds to thousands the next minute.

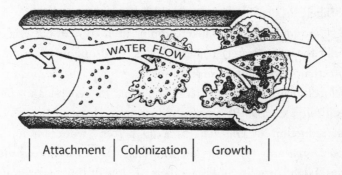

Figure 4.1. Biofilm is a complex mixture of microbes and substances, called a matrix, which adheres to surfaces. A few cells attach to the pipe, then begin multiplying to colonize the surface. In the established biofilm, microbes draw nutrients from the flowing water and store them in the heterogeneous matrix. The biofilm also protects them from disinfectants. *Illustrator, Peter Gaede.*

A glass of tap water holds from less than 100 to well over 10,000 bacteria. Most of them have not been identified because they pose no serious health threat. The great majority of microbes in tap water are harmless. Furthermore, a theory gaining wide acceptance proposes that ingesting water microbes helps contribute to strengthening the immune system, especially in young children.

Damaged and corroded water lines may allow dangerous

microbes into the water system. Fecal contaminants and pathogens then travel quickly throughout the community distribution lines, causing mild to severe outbreaks. *E. coli* and *Salmonella* each survive for at least forty-eight hours in water. Despite our advances in water treatment, the CDC estimates that 900,000 people become sick each year from waterborne diseases. Worldwide, over 2 million deaths annually are attributed to waterborne illness, the equivalent of twenty jumbo jets crashing every day.

Every five years the U.S. EPA (Environmental Protection Agency) issues a list of prevalent microbes in municipal water systems. Bacteria predominate, but parasites (*Giardia* and *Cryptosporidium*), viruses, and algae also show up on the list.

The EPA's current requirements for microbes in drinking water are (1) no more than 500 heterotrophic (a mixture of bacteria commonly found in water) bacteria per cc of drinking water; (2) no more than 5 percent of monthly water samples to be positive for total coliforms, including fecal coliforms in general and *E. coli* specifically; (3) a 99.99 percent reduction in viruses; (4) a 99.9 percent reduction in *Giardia*; (5) a 99 percent reduction in *Cryptosporidium;* and (6) limits on turbidity, or the cloudiness, of the water. High turbidity indicates there are lots of tiny particles in the water that may be carrying microbes.

Each community's microbes are slightly different from those in neighboring towns because of differences in the source waters, hardness, acid levels, and mineral content. We get used to our own town's tap water. That's one reason why people sometimes experience mild to severe discomfort when traveling to different parts of the country.

As the infrastructure of our distribution systems ages—New York City depends on sewer and water lines that are over seventy-five years old—the EPA's list of safety requirements grows. Wells, long thought to be free from risk, may be contaminated with viruses or bacteria when ground seepage infiltrates the well's underground source, called the aquifer. This seepage comes from deteriorating septic systems, excessive rain and runoff, pollutants from overburdened urban systems, or all of these. Septic systems are believed to be the major source of contamination to aquifers, natural springs, and wells.

In these times many wonder about bioterrorism and the safety of their water. Reservoirs are often open to public use and water treatment plants appear easy targets, although they are adopting increased security. Congress currently is considering a number of bills for the protection of public utilities' infrastructure.

Most experts agree that contaminating water may be the most inefficient method of all possible bioterrorism threats. There are three comforting reasons: (1) a biohazard added to reservoirs immediately becomes diluted to the point where even very lethal microbes such as Marburg or Ebola virus do not pose a threat; (2) filtration works well in eliminating pathogens such as anthrax; and (3) chlorine destroys most remaining microbes before they reach the tap. Until, or if, government regulations come into practice for protecting the water supply, the so-called "dilution effect" is our best assurance against lethal waters.

Bottled Water

Bottled water sells for $4 to $10 a gallon. The same volume of tap water rings in at two to three cents. Americans are happy to ignore the downside of their economic decisions to enjoy an expanding assortment of bottled waters. They are marketed as a pure, clean alternative to the stuff coming out of your tap. Imagine a remote stream trickling through alpine forests and filling an aquifer to supply a mountain spring. The water is carefully collected in jugs used for years by the honest local merchants, bottled under strict cleanliness, and shipped to stores. No chemicals, no contaminants, "pure H_2O."

There may be one or two brands that actually do come from pristine mountains. More likely, the bottles are filled straight from a tap. The National Resources Defense Council reports that 25 to 40 percent of commercially sold bottled waters in the United States are "tap water in a bottle." Check the label for "from a municipal source" or "from a community water system." Those phrases mean that the water in the bottle is tap water.

Carbonated bottled water has about the same microbial types and amounts as noncarbonated bottled water. Both generally reflect the quality of the tap water that was put into the bottle.

Bottled water contains roughly the same microbes found in tap water, and sometimes the concentration is greater. The cleanliness of the bottling plant, worker hygiene, and the attention paid to sanitizing reusable bottles all influence the microbiology of bottled water. Bottled water brands follow rules on cleanliness under the authority of FDA. Community utilities that supply tap water follow EPA regulations. The regulations for testing tap water are much stricter than those for bottled water.

"This Water Smells"

Municipal, well, and even bottled water may have off tastes and strange odors. When tap water is unpleasant, people call the local public utility. Customer complaints tend to fall into seven categories: chlorine smell, chlorine smell *and* taste, rotten-egg smell, sewer smell, petroleum (oil) smell or taste, metallic smell or taste, and earthy or fishy smells. Five of these are related to microbiology. Sewer odors come from stagnant water or little-used lines that tend to get a buildup of microorganisms. Rotten-egg smell is from malodorous hydrogen sulfide, a compound produced by anaerobic bacteria hiding in the mud beneath a river, or perhaps at the bottom of the town reservoir. Earthy and fishy symptoms are by-products of intense algae growth. Algae blooms (conditions in which algae multiply rapidly within a certain location) are becoming a serious health hazard in runoff-polluted salt- and freshwaters.

Chlorine smell and taste are associated with water disinfection. Smell and taste together indicate high levels of chlorine in the water. Those living near a treatment plant tend to detect chlorine more often than people living farther away. If water smells like chlorine but doesn't taste of chlorine, there is a surprising reason. Due to the chemistry of chlorine and how it combines with organic compounds, a chlorine smell may mean that *more* of it is needed to balance its chemical reactions. A plant operator will test the water for different forms of chlorine, and may possibly decide to add more until the telltale smell goes away.

Do-It-Yourself Water Treatment

There are instances in which additional treatment of tap water is warranted. Backcountry hikers and campers use filtration units to remove the protozoa cysts, *Cryptosporidium* and *Giardia*. Likewise, people with severely weakened immune systems are advised to filter. In natural disasters and after extremely heavy rains local health authorities announce methods for in-home water purification in case the local source is contaminated. Filtration or bleach is an emergency method for producing water that's safe to drink.

Filtration pitchers and bottles that state on their label a filter pore size of "absolute" one micron are required. They remove cysts and bacteria because each of the filter's one-micron pores is small enough to hold back a three-micron *Cryptosporidium* cyst or a two-micron bacterial cell. Terminology used by some manufacturers can be tricky. The wording "nominal pore size of one micron" means that all the pores *average* one micrometer. In this case, some pores are larger than one micron—big enough perhaps to let cysts and bacteria through. If the label states that the filter is NSF-certified for cyst removal or reduction, or purifies by "reverse osmosis," the device will make the water safe to drink. "Purified water" means that all protozoa cysts have been removed and 99.9 percent of the bacteria have also been removed. Other terms show up on filtration products; accept only those that promise absolute one micron, certification for cyst reduction/removal, or reverse osmosis.

Chlorine tablets or iodine tablets may be substituted for bleach. If using tablets, follow the directions that come in the package.

Amount of Water	Amount of Bleach to Add to Clear Water	Amount of bleach to Add to Cloudy/ Dirty Water
1 gallon	8 drops	16 drops
5 gallons	1/2 teaspoon	1 teaspoon

*** Eight drops is about 1/8 teaspoon; 16 drops is about 1/4 teaspoon ***

Directions: Use unscented bleach (5.25 percent sodium hypochlorite). Shake the solution to mix thoroughly, then let stand thirty minutes before using. Slight chlorine smell should be noticed; if not, add another dose of bleach, shake, and let stand fifteen minutes before using. (Lake Oswego Water Treatment Plant, 2006)

Table 4.1. Bleach is an emergency water treatment.

In the absence of filters and bleach, the CDC recommends heating water at a full rolling boil for more than one minute. Although the CDC recommends boiling as the best method for killing dangerous parasites in suspect waters, most states issue their own "Boil Water Rule" during an emergency. State rules are often more stringent than the CDC's. State boil recommendations range from at least five minutes to at least fifteen minutes. Boiling does not, however, remove toxic chemicals.

Food, Microbes, and You

You probably know more food microbiology than you think. Only three items are needed to perform food microbiology: a stove, an oven, and a refrigerator. Consider yourself an expert if you own a meat thermometer and a refrigerator thermometer. What happens when these conveniences are not available? Each generation has found ways to keep food preserved until it's consumed, and has accomplished it with little or no knowledge of food pathogens.

Foodborne Illness

Microbes and the toxins from microbes and plants afflicted cavemen and cavewomen. The perceptive ones, and survivors who were just plain lucky, shared their knowledge with their children. Over the millennia, cooking and cold storage provided a diet somewhat safe from microbial contamination. Other preservation techniques came about when cooking or cooling wasn't practical or sufficient. Food pathogens adapted to evolving preservation methods and developed wily ways to avoid humans' best efforts to keep food safe and fresh.

Fever, chills, headache, diarrhea, nausea, and vomiting. Sound familiar? Food pathogens cause some of these symptoms. Some deliver *all* of them. Doctors have a hard time diagnosing the responsible organism because many food pathogens cause the same symptoms. The task of following the clues of an outbreak back to the original source of contamination, called a trace-back, is difficult regardless of the microbe's identity. Outbreaks are hard to pinpoint because not everyone gets sick. Not only do the infected tend to move around, but different food pathogens have different lengths of incubation in the body before signs of illness begin. Isolated incidences of foodborne illness therefore often go undetected. (Food-associated illnesses from nonmicrobial causes are referred to as food poisoning. Microbially caused illnesses are called foodborne infection or foodborne illness.) Only when multitudes show up at local hospitals during a concentrated period of time do medical experts suspect foodborne or waterborne disease (or chemical toxicity). In quite a few instances, a sudden increase in demand for diarrhea medicine at drugstores is the clue to an ongoing outbreak.

The Basics of Foodborne Outbreaks

The CDC estimates yearly foodborne illness in the United States may be as high as 76 million cases, most caused by an unidentified contaminant. Of 325,000 yearly hospitalizations, only 20 percent are attributed to a known cause. About twice the fatalities from foodborne illness are never traced back to the contaminant microbe.

To follow the onset of an outbreak, health officials must first determine whether it came from food or water. Microbiologists then identify the new bug, learn its growth requirements, and uncover the details of how it contaminates our food chain. After that, they must figure out the best techniques for preservation. All during that time the new pathogen moves unencumbered through the population, infecting thousands. More often than not the affliction gradually subsides in a community before the mystery is unraveled.

We rarely think about it, but society and lifestyle contribute to the incidence of foodborne illness. Our population seems always to be in the midst of a new infection threatening global health. Furthermore, innovative medical treatments and surgical methods are also unveiled regularly in Western society. Our society therefore includes new pathogens that can infect healthy individuals plus a growing population of those enduring a period of compromised health. Numerous high-risk health situations are on the increase.

There is also increasing evidence that stress, though hard to define, contributes significantly to susceptibility. When the body's defenses are weakened, a door swings open to let disease walk in.

National lifestyle plays a role in pathogen transfer from farm to table. Parents no longer teach children the basics of food harvest, cleaning, and healthy preparation as their ancestors did. The American diet is dominated by packaged foods, loaded with preservatives. Fast foods and restaurant meals are part of virtually every family routine. In other words, we depend on others to provide us with a safe food supply. The food industry is governed by a system of laws, guidances, and standard procedures to ensure food safety. Most of the time, it works.

Populations are gradually migrating toward urban centers. Nowadays, disease can quickly spread to thousands rather than a century ago when perhaps only a few family members got sick because they improperly butchered a hog. Progress comes with risks. In addition, our daily pace continues to increase. Many health experts feel that evolving home and work schedules are contributing to lapses in safe food handling and proper hygiene.

Most of the food-associated outbreaks in history have been related to a microbe; chemicals and pesticides, food additives, and naturally occurring toxins have contributed to a lesser degree.

The CDC lists the following microorganisms as the main causes of foodborne illness: the bacteria *Salmonella, E.coli* and *E. coli* O157:H7, *Staphylococcus, Campylobacter, Clostridium,* and a group of viruses known as Norwalk viruses. Occasional problems are traced to *Shigella* (bacteria), hepatitis A (virus), and *Cryptosporidium* (protozoa cyst). Botulism from *Clostridium botulinum* shows up in rare instances. But when you are rushing to the restroom, you probably don't much care which microbe is the culprit. Unfortunately, at that point it's already too late to worry about prevention.

Microbes in food vary widely. In addition, the number of

microbes that may be in the same type of food can vary. There are also those that cannot yet be identified by microbiologists, so they go undetected until the technology to find them in food improves. The numbers range from less than 10 to over 100,000,000 per gram of food. Food spoilage happens when the microbes reach 10,000,000–100,000,000 per gram. Disagreeable odor, taste, and/or consistency are warning signs. Microbial breakdown of food causes a buildup of end products (bad odor, taste, or color) and degradation of the food's structural components (change in firmness or consistency).

Bacteria that cause spoilage are not necessarily the same as those that carry disease. Pathogens may not cause obvious changes in the food. There is virtually no warning sign for a salad contaminated by a fecal microbe such as *E. coli*. Furthermore, some pathogens need only about a dozen cells to unleash severe illness and sometimes death. Spoilage makes it easy to avoid certain foods because of its warning signs. The contaminants themselves are invisible—a principle of the Five-Second Rule—and may be present in a freshly tossed salad or a sizzling hamburger.

Bacteria thrive in the same foods that provide humans with a rich supply of nutrients—milk, cheeses, meat, fresh fruits and vegetables—and these foods spoil faster than those with a higher level of processing and preservation (frozen or canned). Dry foods, such as nuts and uncooked pasta, have plenty of nutrients but not enough water to support microbial growth. Therefore, dry foods tend to take longer to turn bad and the damage is usually done by molds, which require less moisture to grow.

Raw ground beef, milk, chicken, ham, steaks and roasts, shrimp, green salads, and canned tuna have high numbers of bacteria. Although spices are low-moisture, they are renowned for their high numbers of bacteria and molds. Between 1 and 10 million bacteria are on a gram of ground black pepper; ground ginger holds up to 10 million per gram. This short list of examples doesn't imply that all other foods are microbe-free. Almost everything we eat has bacteria, sometimes in very large amounts.

Food Production Today

The Problem of Contamination

How did our grandparents and their parents manage to put a healthy meal on the table without using a food thermometer and by depending on an icebox of questionable proficiency? The days of growing vegetables in the backyard or slaughtering a hoofed animal once a week are gone. Food production in Western society is characterized by centrally and mass-processed fruits and veggies and massive rendering operations that each day turn thousands of animals into steaks, chops, Buffalo wings, and so on. The efficiency is remarkable. Remarkable, too, is the spread of germs from animal to handler to machinery to hamburger or bag of salad mix.

Vegetables and Fruits

Harvested vegetables and fruits are partially processed in the field. Farm workers remove some leaves and debris before feeding the crop into machinery for squeezing, deskinning, deleafing, etc. The dirt from soil and workers' hands transfers easily to raw produce. Water, air, insects, and fertilizer further contribute to the microbial load. If animals are nearby, their fecal microorganisms also climb aboard. If fields are periodically flooded with waters that have received runoff from cattle operations, they, too, would contain a variety of fecal microbes. Finally, wildlife may deposit contaminants in agricultural fields.

Not only does the plant surface contain bacteria and yeasts, but microbes also move into the plant's vessels. Washing fresh salad vegetables helps, but potentially dangerous microbes remain inside.

Meat and Eggs

The tissues inside an animal's body, such as muscle, nerves, and blood, should be free of all microbes. In other words, they are sterile. (If tissues become contaminated or infected, it can lead to a serious condition called sepsis.) The inside of the intestines, however, contains some of the largest concentrations of microorganisms on earth. In slaughter plants, microbes travel through the air in aerosols, which are very small particles of moisture. Droplets and splashes of intestinal contents easily travel to nearby carcasses. Despite the best efforts in managing airflow and manual hygiene, a slaughter facility is a rough-and-tumble world with rapid and intricate physical actions all around. Contamination of meat is not unexpected under these conditions.

Since microbes and meat are natural bedfellows, civilization has long understood the importance of preserving its source of protein. Meat preservation works by altering the physical environment of the surface, making it inhospitable for bacteria. Refrigeration and freezing, the control of moisture level, and the removal of oxygen in packaging are typical ways to slow contaminant growth. Agricultural colleges spend great efforts in studying optimal handling conditions and packaging materials for reducing potential contamination of our meat supply.

Both steaks and ground beef have a microbial load on their surfaces. But the total surface area of a single steak is considerably less than that of hamburger meat, making hamburger more available as a food source for bacteria. As meat is processed from carcass to large wholesale cuts (chuck, rib, brisket, and shank) to smaller retail cuts (rib eye, top sirloin, short ribs, etc.), the surface area increases. Grinding meat multiplies the surface area enormously; even beef tips for stew have more total surface than roasts and steaks. Meat is nutrient rich. By increasing surface area, more nutrients become available for microbes, and more of the surface is exposed to oxygen. Both conditions help microbes grow.

Eggs are fairly benign as microbial threats. Most eggs are sterile when laid. But they can be contaminated with bacteria such as *Salmonella* or viruses during laying if the hen's ovaries have been infected. Most of the serious consequences of eating raw or undercooked eggs, however, are from microbes transferred from the shell to the contents when the egg is cracked. Like other foods of animal origin, eggs should always be cooked completely to assure that all microorganisms are killed.

The U.S. Department of Agriculture's (USDA) Food Safety and Inspection Service (FSIS) administers programs and inspections to ensure a pathogen-free supply of meat, poultry, and eggs. Currently, the FSIS has established standards for meat and poultry plants to meet when testing for *Salmonella,* generic *E. coli, E. coli* O157:H7 for ground beef and processed meats, *Staphylococcus aureus* toxin, and *Listeria monocytogenes* for ready-to-eat meats, salads, and spreads. Eggs are inspected for *Salmonella.*

Inspecting all domestic and imported meats and eggs is an enormous undertaking and prohibitively expensive. Samples taken for microbiological testing are randomly selected. No surprise therefore, that the program suffers criticism for inadequate or too few inspections. The inspectors don't have superhuman powers of seeing microscopic life. By simply looking at a carcass, they cannot determine the number of pathogens on it. Swabbing followed by laboratory testing has drawbacks. By the time a significant number of carcasses are swab-sampled and checked for microbial growth, the meat has long since gone out on a truck headed for your favorite fast-food spot or grocery. Considering the magnitude of meat production, processing, and consumption in this country, the incidence of pathogen transfer from meat and eggs to humans is very low. Only when an outbreak occurs do we learn of the weak links in the food supply chain.

Organic and Minimally Processed Foods

Organic and minimally processed foods on the farm or in the field contain the same amounts and the same diversity of microbes as foods that are mass produced. Cattle that are certified organic shed *E. coli*, *E. coli* O157, and other pathogens, and these microbes contaminate organic meat exactly as they contaminate nonorganic meat products.

Unpasteurized milk, sometimes sold as "raw milk," is regaining popularity with consumers. Raw milk was the drink of choice several decades ago when families raised their own crops, dairy cows, and meat animals. Though a constant percentage of families had probably become sick from drinking raw milk, the incidences went undocumented. Today, organic raw milk production is on a larger scale and intended for wider distribution. A few hundred people each year become ill from drinking or eating unpasteurized dairy products. The predominant pathogens causing illness are *Listeria monocytogenes*, *E. coli*, *Campylobacter jejuni*, *Clostridium perfringens*, and *Salmonella*.

Pasteurization is done by heating every particle of milk product to a specified temperature for a specified period of time. Fluid milk is pasteurized by heating at 161 degrees F (72 degrees C) for not less than sixteen seconds, or alternatively, 145 degrees F (63 C) for not less than thirty minutes. Mixed dairy products—products that contain milk plus other ingredients, such as chocolate milk, ice cream, or eggnog—receive slightly higher temperatures for the same time periods. Not every microbe is killed during pasteurization, but it has been used for many years without serious outbreaks, confirming that pathogen numbers are reduced to safe levels. The bacteria that remain in pasteurized milk will eventually cause souring. They are *Pseudomonas*, *Alcaligenes*, *Aerobacter*, *Acinetobacter*, and *Flavobacterium*, and all are common in tap water in addition to milk.

The sale of raw milk and dairy products is permitted by certain state laws, but interstate shipment and sale is illegal. Some states do not allow sale of any raw dairy products. If you are a proponent of unpasteurized dairy products, check to be sure suppliers are legally permitted to sell these products in your state. Review information on the dairy's sanitation and inspection records. Some dairies follow exemplary methods for maintaining cleanliness. Beware, however, that no matter how "clean" an operation appears, the microbes remain invisible. The CDC advises against the consumption of any unpasteurized dairy product. The risks are especially serious for young children, the elderly, pregnant women, and immunocompromised persons.

Don't assume organic fruits and vegetables are pathogen-free. First, organic farming uses large amounts of manure in lieu of chemical fertilizers to provide nitrogen to growing plants. Cattle manure spread among the crops greatly increases the chances for *E. coli* O157 and other contamination. Second, organic produce is often from small operations where handling by workers may be more common than on large mechanized farms, increasing chances for pathogen transfer from hands or coughs and sneezes. Third, organically grown products sometimes follow a piecemeal distribution chain. Trucking routes that meander may allow more time for microbes to grow on the produce before it's unloaded at a market.

Consuming organic meats and produce is a choice, and illness rates in consumers of organic products are estimated to be about the same as in those who eat nonorganic foods. Raw dairy foods put certain individuals at risk for illness, and should be avoided.

Probiotic Foods

Probiotic diets are formulated to contain extra amounts of microbes, specifically the lactobacilli and bifidobacteria groups. The idea behind probiotic foods is that current U.S. diets are not properly balanced; they may be high in fats and simple carbohydrates (sugar) and insufficient in complex carbohydrates and protein, and they may contain cancer-causing ingredients among other health hazards. The bacteria added to probiotic foods are expected to deliver several benefits, according to the proponents. Some of the many benefits attributed to the bacterial supplements are competition against pathogens in the gut, prevention of travelers' diarrhea and allergies, enhanced digestion, improved immunity, additional vitamin production, and neutralization of cancer-causing chemicals.

Foods, such as yogurt, that have been on the market for years are a source of the same bacteria added to probiotic foods. At present, there is no clinical evidence that probiotic diets result in the benefits that they advertise.

Seafood

Seafood becomes contaminated in the same ways as meat. Complicated and rapid handling and preparation activities help spread germs, and sloppy hygiene and workers' mistakes on the process line add additional microbes. Exposure to seawater also contributes. Furthermore, fish and shellfish are at the mercy of their native waters. Contaminated fresh and saltwaters lead to contaminated seafood. Oysters, clams, and mussels are most affected because they live in coastal waters where communities sometimes disgorge sewage. Consider that oysters and clams are often eaten raw and the entire animal is eaten, including its intestines. This is one of the reasons why waters that grow shellfish are monitored carefully and are closed when high levels of fecal microbes have been detected. Coastal programs for ensuring safe shellfish are administered by the FDA's National Shellfish Sanitation Program.

Seafood contaminants include the usual suspects: *E. coli, Salmonella,* and *Staphylococcus.* Freshly caught fish are thought to receive most of their bacterial load during onboard processing. Fresh fish should be stored on ice, refrigerated, or frozen until it is used. Never buy any fish that you think has not been stored properly.

Food handling, whether in restaurants, food-processing plants, or at home, has a significant influence on the chances of products being exposed to pathogens. Contaminants straight from fields, farms, or waters are considered rare compared with those from improper food handling. The CDC reports that known cases of foodborne illness are most commonly associated with (a) inadequate holding temperatures, (b) poor personal hygiene, (c) undercooking, and (d) the use of contaminated equipment or utensils, in roughly that order. Bacteria tend to come from all of those methods. Viral contamination of food is rare compared with bacterial, and it is caused almost exclusively by poor personal hygiene by food handlers who are sick. *Cryptosporidium* or *Giardia* contamination is associated with foods from unsafe sources. An "unsafe source" scenario would be, for example, the harvest of fresh vegetables from fields downstream from a dairy farm.

Preventing Foodborne Illness
Of outbreaks from known sources, the most frequently implicated vehicles are fish, shellfish, salads, fruits and vegetables, chicken, and beef, but those from multiple sources are most common of all. The main microbial culprits are *Salmonella, E. coli, Clostridium,* and *Staphylococcus.* Most foodborne pathogens give similar symptoms but have different incubation periods after ingestion before causing illness. They also differ in their infective doses (the number of specific microbial cells needed to make you sick) (Table 4.2).

You have a small measure of control over foodborne infection from restaurant meals and fast food. Don't order very rare or raw meats and seafood unless you're sure the restaurant has an exemplary reputation. Avoid unpasteurized juices and dairy drinks. Glance at your raw salad to assure all the leafy vegetables are washed and clean, free of dirt, sand, or tiny insects. Take note of surroundings: dirty floors, dusty corners, and grimy tables are signs that the kitchen is in

Foodborne Pathogens

Food Illness	Common Foods
Salmonellosis (*Salmonella*)	Raw meats, poultry, eggs, dairy products, fish, shrimp, cake mixes, cream-filled desserts, sauces, salad dressings, peanut butter, cocoa, chocolate
Staphyloccocus Infection (*S. aureus*)	Meat products, poultry and egg products, salads (potato, macaroni), cream-filled pastries, cream pies and éclairs, sandwich fillings, dairy products
Gastroenteritis or traveler's diarrhea (Enterotoxigenic *E. coli*)	Water
Infant diarrhea (Enyero-pathogenic *E. coli*)	Undercooked beef and chicken
Dysentery (Enteroinvasive *E. coli*)	Hamburger, unpasteurized milk
Hemorrhagic Colitis (*E. coli* O157:H7)	Undercooked hamburger, sprouts, fruit juices, ready-to-eat salads
Listeriosis (*Listeria monocytogenes*)	Raw milk, cheeses, raw meat, sausages
Campylobacteriosis (*Campylobacter*)	Improperly cooked chicken
Botulism (*Clostridium botulinum*)	Canned or preserved foods that are not high in acid (green beans, corn, mushrooms, etc.)
Norwalk viruses	Salads, raw oysters, understeamed clams
Hepatitis A	Shellfish, water, salads, cold cuts, fruits and juices, milk products

Table 4.2. Characteristics of foodborne pathogens found in the FDA's Bad Bug Book on the Center for Food Safety and Applied Nutrition Web site.

Incubation Time	Infective Symptoms	Dose
6–48 hours	Nausea, vomiting, abdominal cramps, diarrhea, fever, headache	15-20 cells
Rapid onset	Nausea, vomiting, retching, abdominal cramps, prostration	Less than 10 micrograms of toxin, or over 100,000 bacteria per gram of food
24 hours	Watery diarrhea, cramps, low-grade fever, nausea, malaise	100 million to 10 billion
24 hours	Watery or bloody diarrhea	Less than 100 cells for infants; 1 million for adults
24 hours	Blood and mucus in stools	10 cells
24 hours	Severe cramps and bloody diarrhea	About 10 cells
12–24 hours	Septicemia, meningitis, encephalitis	1,000 cells or less
2–5 days	Diarrhea, muscle pain, headache, abdominal pain, fever	400-500 cells
18–36 hours	Weakness, vertigo, double vision, difficulty breathing and swallowing	A few nanograms of toxin
24–48 hours	Nausea, vomiting, diarrhea, abdominal pain (mild and brief illness)	About 10 viruses
10–50 days depending on dose	Mild fever, malaise, nausea, anorexia, abdominal discomfort, jaundice	Less than 10 to 100 viruses

Center for Food Safety and Applied Nutrition Web site.

no better shape. In fast-food places it's easy enough to take a minute to watch the behavior and hygiene practices of the servers at the counter. Sometimes the food-prep area is visible to customers. Check if workers wear gloves and hairnets; watch to see if they remove gloves as they leave the area and put on new ones when they return.

Gloves for handling food are effective only if they are changed. New gloves should be used whenever returning to the food-prep station. Used gloves are to be discarded as soon as a specific job is complete, such as handling raw hamburger patties. The next time you're in a fast-food restaurant, take a few minutes to watch all the items the handlers touch, all the activities they conduct, while wearing the same pair of gloves.

Food Safety Basics

PRACTICE GOOD HYGIENE Pull back hair; wear clean clothing or apron; don't prepare meals when sick

WASH HANDS Follow instructions in Chapter 3
• Before preparing meals and after each break, after removing trash, and after handling money
• Immediately after handling raw meat or fish, *each* time

COOK AT THE PROPER TEMPERATURES Use an accurate thermometer placed at the center of the food; cook until temperature reaches at least:

Poultry (whole or ground), casseroles,

stuffed meats: 165 degrees F

Hamburgers, ground beef, pork, or lamb: 155 degrees F

Pork: 150 degrees F

Beef, lamb, seafood, eggs: 140 degrees F

Beans, rice, pasta, potatoes: 140 degrees F

Rare roast beef: 130 degrees F

After all your best efforts to assure a safe meal, you have no recourse against the fellow back in the kitchen touching his nose and mouth before thrusting his hands into the salad greens. Accept that almost all of us will fall victim from time to time to a mild or ferocious case of foodborne illness.

At home you have more control over preventing foodborne disease, and there are few excuses for getting sick, assuming you own the required equipment: stove, oven, refrigerator (and freezer), thermometer. But foodborne illness does occur from home-prepared meals. People take shortcuts or forget good hygiene. They're too busy or confident that "it won't happen to me." Human nature tips the scales in favor of food pathogens (Table 4.3).

COOL LEFTOVERS QUICKLY AND COMPLETELY Reduce the time that food remains in the "Danger Zone," between 45 and 140 degrees F; refrigerate within four hours; do not cover until cold and do not stack pans when cooling; cut meat into 4-pound chunks or smaller; stir liquids (soups, sauces) frequently while cooling

REDUCE CHANCES OF CROSS-CONTAMINATION In refrigerator, store ready-to-serve foods on top shelves, raw vegetables in the middle, and raw meats on the lower shelves; prevent thawing foods from dripping on other foods; clean and sanitize utensils and cutting boards before use; avoid using the same cloth towel repeatedly for drying hands and wiping down surfaces; throw away any raw vegetables that have touched raw meat, poultry, seafood, or their juices

Table 4.3. Food safety tips from the Environmental Health and Safety Department, Washington State University

The FSIS publishes helpful fact sheets on the handling and preparation of the foods regulated by the USDA: meat; poultry; eggs and egg products; seasonal foods such as barbecues, holiday meals, and camping/hiking food; and mail-order foods.

The rules of thumb to follow in preventing foodborne illness are: cook meats until the juices run clear; cook fish until it flakes; cook eggs until they are no longer runny; wash fruits and vegetables thoroughly with cold running water; assure that the quality of your community's tap water is good before using it to wash foods; and be diligent in watching out for cross-contamination among cutting boards, utensils, and other surfaces. Cross-contamination is the transfer of live microorganisms from uncooked foods to other foods. A common method of cross-contamination is placing vegetables on a cutting board that had just been used to cut raw meat.

E. coli O157 causes sporadic serious outbreaks in fresh salad mixes. Even if a mix is labeled "washed" or "ready-to-eat," wash all salad mixes and green leafy vegetables. Always wash hands properly before handling raw vegetables and fruit, and do not allow them to be recontaminated by coming into contact with raw meat.

Scientists disagree on whether foodborne disease is more of a hazard today than in years past. Modern methods in food handling, processing, and production-plant sanitation have reduced many serious threats. But mass production and imports are on the rise. Meal preparation in the home has decreased, as has an emphasis on hygiene. Every day new pathogens emerge in our food supply chain.

Defenses against Pathogens

Some foods provide good conditions for pathogens to grow in, and some offer only a harsh environment. Foods that have high amounts of acid ward off invasion because many bacteria cannot tolerate a low pH (i.e., acidic conditions). A few microbes are able to grow in acidic environments, and these are the ones that must be managed carefully with

High Acid Foods—Grapefruit, lemons, limes, oranges, peaches, pineapple, plums, berries, cherries, grapes, apricots, apples, tomatoes, juices, jams/jellies, vinegar, pickles, green olives, sauerkraut, buttermilk, mayonnaise, and soft drinks.

Acidic Foods—String beans, beets, brussels sprouts, carrots, lettuce, onions, parsley, peas, peppers, potatoes, spinach, squash, melons, bananas, ground beef, ham, oysters, salmon, most cheeses, butter, cream, egg yolk, and bread.

Neutral Foods—Corn, cantaloupe, lamb, pork, chicken, turkey, fresh fish, shrimp, crabs, clams, and milk.

Basic (Nonacidic) Foods—Camembert cheese, egg whites, whole egg, crackers, and most cakes.

preservation. Dry foods and low-moisture foods are also difficult places for most pathogens; only microbes that tolerate low moisture levels—that is, molds—can cause spoilage in dry conditions.

Bacteria that do well in acidic foods are *Campylobacter* and *Lactobacillus*. Some are comfortable in a range from acidic to neutral: *Salmonella, S. aureus, C. botulinum, Listeria monocytogenes, Streptococcus pyogenes,* and *Bacillus cereus. Shigella* is one of the few food pathogens that prefers nonacidic foods.

Processing methods affect food's natural defenses. **Drying** is the oldest means of preservation. Besides inhibiting microbial growth, it also prevents deterioration and makes storage and packaging easier. **Smoking** is another ancient method. Chemical compounds created from wood smoke are antimicrobial and also give food a distinct flavor. **Curing** uses compounds that change the flavor, color, or texture of food, while inhibiting spoilage. Traditional curing methods include high sucrose (sugar) levels, high sodium chloride (salt) levels, and addition of sodium nitrite or sodium nitrate. The nitrites and nitrates are used in hams and bacon.

Physically altering food helps prevent bacteria and molds from growing at their optimal conditions. Fresh fruit is stored in slightly reduced oxygen levels with a corresponding increase in carbon dioxide. This method is used also for some meats and fish. Vacuum packaging also serves the purpose of reducing the food's exposure to oxygen.

Preservatives

When we think of a preservative, it is usually a chemical added to food—a chemical that would not normally be found in nature. But sugar and salt—chemicals found in nature—are also preservatives. Organic acids are found in nature as well and are added to foods because of their antimicrobial benefits. Sorbic, benzoic, acetic, and citric acids make certain foods inhospitable for microbes, and at the same time they give food a distinct taste (sauerkraut, pickles, vinegar).

Common Food Preservatives

Work Best against Bacteria:

Sodium or potassium chloride	Sodium ascorbate
Lactic acid	Ascorbic acid (vitamin C)
Citric acid	Sodium erythorbate
Sodium nitrite, sodium nitrate	Butylated hydroxyanisole (BHA)
Propylene glycol	Butylated hydroxy-toluene (BHT)
Ethylenediaminetetra-acetic acid (EDTA)	

Other chemical preservatives sound downright sinister or incomprehensible. It may give you pause to notice sodium hexametaphospate, potassium bisulfite, or ethylenediaminetetraacetic acid written on a food packaging label. The FDA approves preservatives in one of two ways. Many are tested in animals and considered safe in the levels consumed by humans. Others have not been tested extensively but have been used for so many years that the government believes there is enough historical evidence to consider them safe. These are referred to as **GRAS** compounds—Generally Recognized As Safe. For many people, there remains a balancing act in choosing between the potentially deleterious effects of chemicals consumed over a lifetime and the devastating immediate effects of a deadly food pathogen.

Acids change the properties of food, thereby inhibiting the growth of bacteria and molds. Acids also interfere with membranes, enzymes, and nutrient uptake by the microbe. Sugars soak up water, making it unavailable for use by bacteria, and salt destroys cell membranes, enzymes, and other cellular activities by releasing ions.

Work Best against Molds:
Benzoic acid   Potassium sorbate
(mostly against yeast)
Sodium or potassium benzoate (mostly against yeast)

Inhibits Both Bacteria and Molds:
Sodium phosphate   Sodium dehydroacetic acid
Propionic acid   Sodium aluminum
                                          phosphate
Calcium propionate   Phosphoric acid
                                          (soda drinks only)
Sodium bisulfite

Nitrites and nitrates are compounds known as oxidizing agents, which have chemical properties effective against anaerobic bacteria. Ethylenediaminetetraacetic acid (EDTA) doesn't affect the microbe directly but makes other preservatives more powerful.

Food preservatives work best when added in combination. No single preservative can meet every requirement for inhibiting microorganisms, blending with the food's ingredients, and imparting no effect on taste and odor. Preservative combinations are selected based on the type of food and the types of microbes most likely to cause spoilage or contamination. When used together these compounds are synergistic; that is, in combination each preservative has more effect than it would if used alone. In foods this enhanced preservative activity is called the hurdle effect (also the barrier effect). The idea is that a multitude of preservatives force each microbe to bypass various challenges before growing in and spoiling the food. One preservative may only

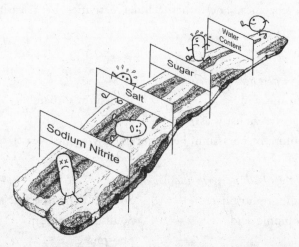

Figure 4.2. The hurdle effect in a piece of bacon. Salt, sugar, water content, nitrates and nitrites, and even packaging that keeps out oxygen force microbes to hurdle a series of preservatives before they spoil the food. (Adapted from M. R. Adams and M. O. Moss, 2000, *Food Microbiology*, 2d ed., Cambridge, UK, Royal Society of Chemistry.) *Illustrator, Peter Gaede.*

weaken the bug; the second damages the weakened microbe some more; the third easily kills the microbe made vulnerable by the other preservatives. Due to the hurdle effect (Figure 4.2), each preservative may be added in amounts less than would be needed if just one preservative were used.

Beneficial Food Bacteria

There are a variety of bacteria, yeasts, and molds present in food for a reason. Some give a distinctive taste (pickling), some transform one type of food to another (fruit to wine), and some are added to enhance preservation (cheese).

The techniques used in adding select microbes to certain foods have been practiced for thousands of years. These new foods increased variety in the diet of early societies but also allowed people to utilize their food supplies year-round without losing them to spoilage. This may be considered the first true practice of bioengineered foods. For example, *Saccharomyces carlsbergensis* yeast (bio-) ferments the amber liquid called hopped wort, which results from the proper cooking, steeping, and mashing (engineering) of barley, to produce beer.

One goal of bioengineering is to increase the efficiency of worldwide food production. Bacteria, yeasts, and viruses have been used for this purpose since the mid-1990s. In this type of technology a gene containing the instructions for a desirable characteristic (referred to as a trait) is identified in a microorganism, copied, and then inserted into the genetic material (DNA) of a plant or animal. The insertion may take place by a number of methods. One common method is to put the beneficial gene into another microbe that is known to infect the target plant cells or animal tissue (Figure 4.3).

Some varieties of corn, rice, and soybeans are bioengineered crops. Their new attributes include increased defense against pests,

Microbes and Food Production

Start with . . .	Add . . .	Get final product . . .
Cabbage	Lactic acid–producing bacteria (*Lactobacillus, Leuconostoc, Streptococcus, Pediococcus*)	Sauerkraut, kimchi
Cucumbers		Pickles
Beef		Sausages and bologna
Pork hams	*Aspergillus* and *Penicillium* molds	Country-cured ham
Flour	*Saccharomyces cerevisiae* yeast	Bread
	Lactobacillus sanfrancisco bacteria and Candida *humilis* or *milleri* yeast	Sourdough bread
Soybeans	*Saccharomyces* yeast	Soy sauce
Molasses, sugar cane	*Saccharomyces* and *Schizosachharomyces* yeasts	Rum
Fruit, fruit juices	*Saccharomyces yeast* and *Oenococcus* bacteria	Wine
Corn, rye	*Saccharomyces cerevisiae*	Bourbon whisky
Rice	*Saccharomyces saki*	Sake

resistance to pesticides that kill other plants, and ability to tolerate extreme weather. MacGregor tomatoes contain a microbial gene that prolongs shelf life. In food animals, salmon and shrimp are engineered to resist disease, use nutrients efficiently for growth, and withstand water temperatures outside their normal optimal range. Other focus areas of biotechnology are higher yields or growth rates, immune system enhancement, host-produced pesticides, and increased vitamin content. Removal of food's natural toxins and allergens are also of interest.

Biotechnology's workhorse is *Bacillus thuringiensis,* nicknamed

Start with...	Add...	Get final product...
Apples	*Saccharomyces* yeast	Cider
Cider, wine	*Acetobacter* bacteria	Vinegar
Milk	*Lactobacillus*, *Streptococcus*, and *Bifidus* bacteria	Yogurt
	Lactic acid bacteria	Cottage cheese, cream cheese
	Streptococcus bacteria	Cheddar cheese
	Streptococcus, *Lactobacillus*, and *Propionibacterium* bacteria	Swiss cheese
Cream	*Streptococcus* bacteria buttermilk	Sour cream,
Unripened cheese	*Penicillium roquefortii* mold	Roquefort, Stilton, blue, Gorgonzola cheeses
	Penicillium camemberti mold	Camembert cheese

Bt. This bacterial species is common in soil. Its value comes from the ability to produce an insecticide that protects corn against corn borer caterpillars, potatoes against the Colorado potato beetle, and tomatoes against corn earworm. Farmers have for decades sprayed the Bt insecticide, or Bt toxin, directly onto their crops. A more sophisticated approach involves transfer of the gene for the toxin from *B. thurengiensis* into another microbe that invades the plant and thus puts the new gene into the plant's DNA. Viruses have the most potential in serving as the gene's carrier.

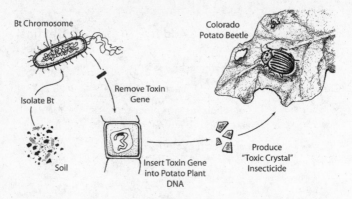

Figure 4.3. Potato plants are bioengineered to be resistant to damage from the Colorado potato beetle. *Bacillus thurengiensis* (Bt) bacteria make natural insecticide crystals that are toxic to the beetle. Bt is isolated, then the gene for the insecticide is located in its chromosome. Using gene transfer, the gene for the toxic crystals is put into potato plant DNA. The plants can now produce the insecticide on their own. *Illustrator, Peter Gaede.*

Controversy surrounds bioengineering, mainly because it involves unpredictable biological systems. In agriculture, these systems are further affected by irregular weather patterns and climate. Many feel there are critical ethical issues in devising "unnatural" entities with special characteristics they would not naturally possess. Safety for humans and ecosystems are also topics being debated. Like the discovery of DNA itself, the ability to manipulate DNA to contain new genes was a milestone in the advancement of science. The unknown is always accompanied by questions, argument, fear, and new discoveries.

Here is a list of some of the current concerns regarding genetically engineered microbes:

- New species that are toxic to humans
- Creation of new allergens that affect human health
- Long-term impact on ecological systems and their native life-forms

- Release of engineered species to neighboring farms and fisheries, or into nature
- Release of previously unknown harmful biological activities
- Disruption of the normal distribution of genes on earth
- Loss of biodiversity
- Proliferation of transferred genes causing imbalance in plant and animal ecosystems
- The ethics of genetically manipulating life
- Possibility of using bioengineering techniques for bioterrorism
- The potential for large companies to control the world's food supply with bioengineered products

Summary

In food and water microbiology, tap water and our food supply may be safer than they have ever been. But there is evidence that the global distribution of pathogens in food is more extensive than ever. Microbes adapt to human inventions faster than humans adapt to them. For that reason alone, we likely will never see a day in which the world is free of pathogens, including those in food and water. It's unlikely, too, that each major step in biology will be met with unreserved acceptance. The science of destroying deadly microbes and the technology of using beneficial ones will continue to receive debate and introspection. These intellectual labors are probably the greatest contribution humans can make to life among the microbes.

The Famous and the Infamous

"Happy families are all alike; every unhappy family is unhappy in its own way."

Leo Tolstoy probably didn't realize how accurately he described the world of food pathogens. Most bacteria live in harmony with humans, but food pathogens do not. Gram-positive *Clostridium perfringens* and Gram-negative *E. coli* O157:H7 are examples. *C. perfringens* is common in nature; its spores survive for years in soil. *E. coli* O157 is a tenderfoot by comparison, living only in animal intestines and traveling to new hosts by way of fecal contamination.

C. perfringens causes mild-to-moderate illness after ingestion of millions of cells in improperly heated or cooled foods. Stews, gravy, sauces, and casseroles are common. Cooking kills most of the spores, but those that are not killed may actually become activated by the heat of cooking. When bacteria go from their spore form to their reproductive form, they release a powerful toxin, which causes severe intestinal irritation. Diarrhea and abdominal cramping start 9 to 15 hours after ingestion and last about 24 hours. *C. perfringens* is an anaerobe, but it can tolerate exposure to air. It prefers locations like the oxygen-less intestines. Though it receives little publicity, "perfringens poisoning" is one of the most common foodborne infections. Some health specialists believe that the true number of cases from *C. perfringens* is 20 to 30 percent higher than the estimated 250,000 cases/year.

E. coli O157 outbreaks are rare compared with those of *C. perfringens*, but they get attention because they often include deaths, and because each outbreak spotlights the hazards of our current methods of mass food production. Foods associated with outbreaks are raw and undercooked ground beef and hamburger, sausage, alfalfa sprouts, green leafy vegetables, unpasteurized fruit juices and milk, and some cheeses. In all cases, the food had been contaminated with fecal matter. Cross-contamination during meat processing and in-field or in-process contamination of crops are the reasons O157 ends up in food.

As few as ten O157 cells cause sickness, which is characterized by severe, often bloody, diarrhea and cramping that last for an average of eight days. O157 and other *E. coli* strains do not form a protective spore coat and do not naturally live outside the intestines. *E. coli*'s metabolism is opposite that of *Clostridium*. It's a facultative anaerobe that can live without oxygen but grows best in the open air. The elderly and young children and infants suffer serious complications, that is, neurological symptoms or kidney failure. The time to onset of symptoms varies (two to nine days are commonly reported). O157 outbreaks have been traced back to foods at nursing homes, county fairs, state parks, sports camps, and beaches. Undoubtedly, there are dozens of other places where the microbe can be found.

O157 cases are easier to follow than *C. pefringens* illnesses. That's because O157 symptoms are severe enough to warrant a trip to a doctor. *E. coli* O157 cases are estimated at about 73,000 per year. *C. perfringens* causes less havoc in the body, so thousands of illnesses are unreported and unnoticed by the health community. In addition, U.S. health agencies monitor O157. *C. pefringens* is not monitored, and doctors report it only on a voluntary basis.

C. perfringens and *E. coli* O157 are examples of the remarkable diversity of food pathogens. They are examples, too, of the quirks that make one microbe a media darling and another a veritable unknown.

5

Clean It Like You Mean It

I'm gonna wash that man right outa my hair.

—Oscar Hammerstein

Within the past five years, Americans have spent more than $1.6 billion on household cleaning products. The spending continues, perhaps revealing the current state of our collective lifestyles. Many households have two wage earners with little time for housecleaning, so the products they buy had better be first-rate. An almost constant stream of news stories have microbiological themes: faulty flu vaccines, outbreaks, pandemic diseases, and biological weapons. There seem to be daily discoveries of new antibiotic-resistant "superbugs." In all these instances, industry is willing to admit that fear is a critical factor, fueling the increased sales of disinfectants and sanitizers.

Chlorine was discovered in 1774 and used in the early nineteenth century for reducing odors in ships, prisons, stables, sewers, and countless other malodorous locales. With the odor controlled, people incorrectly assumed that they had also stamped out contagious disease. Gradually the principles behind microbe transmission became understood, and experiments confirmed that microbes cause infectious disease. Aided by emerging refinements in chemistry, microbiologists found the ways to kill pathogens with chemicals, as well as by cooking, salting, and smoking. By the early 1900s, disease control meant the use of phenols, hypochlorites, carbolic acid, alcohols, and hydrogen peroxide. These chemicals remain in use today, plus a variety of synthetic compounds, all for the purpose of defeating pathogens that have plagued humans throughout history.

Along with concern over epidemics, society puts increasing emphasis on protection of the environment. The effects of chemicals on water, earth, and air are viewed with greater urgency. The can of kitchen cleaner, nestled in the cabinet between sponges and the mop, has become either a key to good health or the harbinger of ecological doom. Nowadays the proliferation of disinfectants has created a tangle of questions and viewpoints. Who would have thought the innocent-looking bottles in the grocery aisle would be so controversial?

Disinfectants are effective and recommended for killing bacteria and viruses when juices from raw chicken spill onto the kitchen counter or a flu sufferer sneezes directly onto the phone you are about to use. But how useful is a can of antimicrobial spray in the hands of a healthy person practicing good hygiene? In other words, "How clean do we need to be?"

There are instances beyond handling raw hamburger and chicken, sitting next to a sneezing and drippy commuter, or venturing into the airport restroom, where extra care and cleaning is warranted. Portions of the population are considered permanently at risk or in a temporary high-risk health situation. Examples include newborns and the elderly, pregnant women, AIDS patients and others with weakened immune systems such as cancer patients and transplant patients, those with diabetes, children in day care, and residents of nursing homes.

The numbers of children in day care and the size of the over-sixty population are increasing. The average length in days of a hospital stay is being reduced. Advances in medicine are saving more lives among the critically injured or diseased, yet the ironic result is an increase in the number of people requiring special medical care. Innovative organ transplantations and new chemotherapies for cancers add to the subpopulation of people with increased susceptibility to infection. In these instances, the use of antimicrobial products is vital.

Antimicrobial Basics

An antimicrobial substance, natural or synthetic, kills or suppresses the growth of microbes. Antimicrobial products act on bacteria, molds, yeasts, mildew, algae, protozoa, and viruses.

In the medical world, the term "antimicrobial" refers to drugs applied in or on the body to kill microbes: antiseptics, antibiotics, and chemotherapeutic agents. The FDA is responsible for supervising the antimicrobial formulas used in human and veterinary medicine.

In household cleaning, an antimicrobial substance is a product or active ingredient in a product that kills microbes. The EPA is in charge of governing these products. In the early 1970s to mid-1980s, the EPA began ensuring that all pesticides are safe and effective, and antimicrobial products were categorized as pesticides, along with things that kill insects, snails, worms, and rodents.

A **biocide** is a product or chemical that kills any living organism from insects and rodents to microorganisms. Disinfectants, sanitizers, antiseptics, and preservatives are sometimes referred to as biocides. Antibiotics are also biocides, but are classified as drugs. **Germicide** is another term used for products that kill microbes.

Disinfectants and sanitizers are two types of antimicrobial products. Within the disinfectant group, there are additional specialties: bactericides kill bacteria, fungicides kill fungi (molds including yeast and mildew), virucides kill viruses, algicides kill algae, and so on. Sanitizers target bacteria; they are not for viruses, molds, or other nonbacterial microbes.

A product is called a **disinfectant** only if it has been shown to kill, within ten minutes, at least a million bacteria or fungi on a hard inanimate surface. The testing is done on one microbial species at a time. The criteria for virucides, the disinfectants against viruses, are slightly different due to the different way viruses are grown in a laboratory. At least one spray disinfectant sold today is especially effective because it kills several bacteria, several viruses, and a fungus all at once.

Disinfectants are for hard surfaces such as counters, sinks, toilet seats, showers, floors, tables, hospital gurneys, operating tables, plastic or metal garbage cans, and so on. A few disinfectants work on porous surfaces, too, such as tile grout and wood. For those, the product label states that the formula works on porous items. Products for use on laundry and fabrics (upholstery, curtains, carpets, etc.) are in their own categories separate from hard-surface disinfectants.

Disinfectants are not for use on humans or animals. The terminology can become confusing. Sadly, even the World Health Organization refers to alcohol as a "skin disinfectant," an incorrect use of the term.

A **sanitizer** reduces the number of bacteria to *safe levels* as defined by the EPA. On household surfaces, laundry, carpets, and in indoor air, a reduction of 99.9 percent of bacteria within five minutes is considered a safe level. On equipment or utensils used in food preparation, "safe" is a reduction of bacteria by 99.999 percent within thirty seconds.

How safe is "safe"? If you start with 1,000 bacteria, killing 99.9 percent of them leaves you with one live bacterial cell. Killing 99.999 percent of 100,000 bacteria leaves you with one cell. What if the numbers of bacteria are a million per droplet of raw hamburger juice? A sanitizer that eliminates 99.9 percent of the million, still leaves 1,000 live bacteria. (Understand that these numbers are estimated rather than exact. It's virtually impossible to kill exactly 999 bacterial cells in lab tests.) That might be okay for killing *Salmonella,* which needs a dose of 100 to 1,000 cells to make you sick, but *E. coli* and *Shigella* are dangerous if you swallow as few as ten cells. Moreover, you have no way of knowing just how many bacteria are in that droplet on your kitchen counter. This is why disinfectants are more powerful than sanitizers in eliminating any bacteria that may be lurking.

The flip side to the question "How safe is safe?" is the question "How clean do we need to be?" Are a million or more bacteria really sitting on the counter, your keyboard, or a toilet seat? Has a hanky, fresh from the dryer, ever made you nauseous? Bacteria are everywhere in your home. Sometimes their concentration is high, sometimes low. Perhaps a disinfectant that wipes out over a million microbes merely illustrates a case of "overkill" because the overwhelming majority of microbes on earth are not pathogenic. Therefore, the odds are in your favor when you eat a Five-Second Rule cookie.

Cleaning

There is dirt and there is filth. Dirt is not always visible. It includes dry or moist substances such as juices from meat and vegetables, mucus and saliva, and tiny droplets of milk, blood, or urine. Dirt also includes small bits of garden soil, animal dander, feces, soap film, and dust or other particles carried in the air. Filth is the dirt you see. It includes all the above, but in large quantities.

For disinfectants to work, dirt and filth must first be removed (Figure 5.1). Dirt makes germicides less effective because microbes hide in tiny nooks and crannies or use dirt's proteins to protect them. Therefore, cleaning (sweeping, mopping, wiping with a clean sponge or towel) before applying the disinfectant helps ensure the product has killed the microbes it's supposed to kill.

Certain products are called "disinfectant cleaners." These simultaneously remove dirt *and* disinfect. This is one-step cleaning. Most of the popular disinfectants and sanitizers sold are not one-step cleaners. Their labels include specific instructions for removing dirt—look for the term "precleaned surface"—BEFORE using

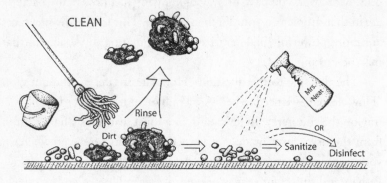

Figure 5.1. Dirt should be removed before using most disinfectants and sanitizers. Fortunately, the dirt pictured is not to scale! *Illustrator, Peter Gaede.*

them. This is two-step cleaning. Sometimes a label will state that only "heavily soiled surfaces" need to be precleaned. It's up to you to decide if you are dealing with a *heavily* soiled surface.

The product label is important because it describes how the product must be used in order to destroy the target microbes. But reading a label is one thing; following its instructions is another. Most people don't follow the directions. Though a disinfectant may require five or even ten minutes to work, it's unlikely most people wait. Your household arsenal of sprays, wipes, and solutions need time to do their job. The active ingredient must infiltrate the microbe's cell wall and membrane before it disrupts normal cell functions. This doesn't happen instantaneously. The period needed for a product to kill microbes is called **contact time**. If a product says it "kills bacteria instantly," don't believe it. Even the strongest disinfectants, like bleach, require contact time. Sanitizers' contact time is shorter than those of disinfectants, from thirty seconds to five minutes, because their requirements are less stringent.

Makers of disinfectants and sanitizers compose marketing claims that promote all the plusses of their products. Antimicrobial

Figure 5.2. (a) Disinfectants must give directions for how to use them, the types of surfaces they disinfect, and the microbes they kill. (b) This product has been shown to kill three different bacteria, which represent many other household bacteria. When all three are listed, it means the product may be used in hospitals and medical clinics in addition to the home. (c) The label must include precautionary statements regarding health hazards and the proper way to dispose of the empty container. *Courtesy of Prestige Brands. Copyright 2006 Spic and Span Company/Prestige Brands Inc.*

claims may sound like nothing more than catchy marketing slogans, but they are carefully crafted phrases that draw a shopper's eye *and* meet the EPA's requirements.

The EPA looks for three main points before approving a product: (1) the directions for using it, including contact time; (2) the names of the microbes the product kills or inhibits; and (3) the active ingredients that do the killing or inhibiting (Figures 5.2 and 5.3). Even antiviral facial tissues that claim to kill "99.9 percent of cold and flu viruses" are registered with the EPA and contain all the required information on their label.

Germicide manufacturers occasionally claim that they promote a "healthy home" or a "healthy family." This has never been proven, and it's not likely to be. Germs enter your house on your shoes, on food, and on your hands. They blow in through open windows and ride in on your pets within minutes after you disinfect a kitchen or bathroom. Disinfection in your home does not protect you from

Figure 5.3. This product disinfects (contact time is 4 minutes) or sanitizes (contact time is 30 seconds). (a) Antimicrobial products must list their ingredients, identifying the active ingredients. (b) A telephone number is included for consumers with questions or comments. (c) The registration number must be on the label. All of the information must be legible. *Courtesy of The Clorox Company. Copyright 2001, 2003, 2004 The Clorox Company.*

catching a cold from the guy standing behind you in the checkout line. Though antimicrobial products are important in certain situations, your home or family will not become healthier if you use them.

Antimicrobials and Microbial Resistance

Microbiology is not without its controversies. One of them is the debate over the effects, if any, of chemical cleaners on the rise of antibiotic- or chemical-resistant microbes.

Resistance is the ability of a microbe to block the action of an antimicrobial compound, such as an antibiotic. Bacteria and viruses develop resistance through chance mutations during replication. In bacteria, occasional mutations result in genes that direct the cell to make new enzymes. Some of these enzymes destroy antibiotics.

Bacteria have traits that help them quickly develop resistance not only to antibiotics but also to other threats. Many bacterial generations develop in a short span of time, and through each generation individual cells have a chance to pick up and retain characteristics that are helpful to their survival. Some may develop a talent for living in very high temperatures or salty solutions. Others form mechanisms for resisting dangerous chemicals, such as benzene or arsenic or an antibiotic. The few cells that by luck retain a gene for an antibiotic-destroying enzyme will have a huge advantage over other, more defenseless, cells.

Antibiotics are substances made by fungi or bacteria or synthesized in laboratories that kill other microbes. The antibiotic penicillin was introduced in the 1940s and was soon prescribed for various ailments, whether or not it was the right choice. In the intervening years, an increasingly diverse collection of bacteria acquired traits for fighting off the effects of penicillin. Most of these bacteria have since developed resistance against additional antibiotics. Resistant bacteria are now recognized as one of the major problems in health care, and they came about due to years of unrestrained use of antibiotics. Today the medical community and microbiologists are worried that antibiotic resistance is outpacing our capacity to invent new drugs to treat infections.

Another bacterial attribute is the **plasmid**. A plasmid is a small string of DNA that floats free in the watery contents of the bacterial cell. The building blocks of DNA (nucleotides) are connected in a precise order. Sections of DNA's ordered nucleotides make up genes, and it is the plasmid's genes that give the cell instructions on how to be resistant.

The cell's main storage of DNA and genes is the chromosome. Chromosomes are larger than plasmids and they, too, hold instructions for resistance. Compared with chromosomes, plasmids have the advantage of being small and handy to pass around among friends. It is this transfer of plasmids between cells of the same species, or sometimes between different species, that helps bacteria become impervious to drug treatment.

Antibiotic resistance has been accepted as a real and difficult problem for several years. Resistance to the antimicrobial chemicals found in cleaning products, however, is not universally accepted. The liberal use of cleaners, disinfectants, and sanitizers has convinced some microbiologists that these products lead to chemical-resistant bacteria. If resistance to chemicals can be induced, the argument follows that chemical-resistant bacteria will use similar means to become antibiotic resistant, thus feeding the impending disaster of microbes that cannot be killed by any known drug.

Others argue that the action of chemicals is not related to the activity of antibiotics. Chemical disinfectants, as the argument goes, simply wipe out any bacteria in their path; there's no chance for resistance to emerge. The controversy continues to capture headlines.

The Case *For* Disinfectants and Sanitizers

Proponents of chemical disinfectants offer a number of strong arguments in favor of regular disinfection and sanitization. Some microbiologists and nonscientists propose that these products are important for use in all homes, restaurants, office buildings, and anywhere people congregate, not just hospitals. Some of their arguments are presented here.

Disinfectants and sanitizers are essential for destroying disease-causing microbes in homes, hospitals, day care centers, nursing homes, hotels, restaurants, and public restrooms, among other places where infection spreads. They eliminate pathogens from medical and food processing equipment and make tap water safe to drink. Germicides

The Case *Against* Disinfectants and Sanitizers

Opponents of chemical disinfectants offer a number of strong arguments in favor of their prudent use. They propose that these products are necessary only in certain situations as a precaution against infection and should not be used to wipe out the normal and harmless microbes all around us. Some of their arguments are presented here.

Disinfectants and sanitizers are ineffective because they do not work as advertised when you don't follow the directions, and people very rarely follow them. Live bacteria remain on the "disinfected" surface when you spray the product and immediately wipe it away. Sanitizers are even worse. They're intentionally formulated to leave behind a small percentage of live bugs. Only the strongest "superbugs" remain, and as combined with good personal hygiene are more important than ever as the number and size of at-risk groups continue to grow. Studies on the in-home use of disinfectants show that, when used along with regular cleaning, they reduce numbers of bacteria that could cause infection.

Because *E. coli, Salmonella, Campylobacter, Staphylococcus,* and others have all been found in typical household kitchens, other pathogens are likely to be found there, too. You cannot see them so you don't know when and where they are present. Germicides are therefore necessary in reducing the threat of disease being transmitted in your home.

These products do not contribute to resistance. They have been used for years with no significant increase in chemical resistance. Resistance develops over several cell genera-tions, each with an extremely small number of spontaneous mutations. The fastest mutation rate (the chance that a gene will mutate each time a cell divides) in bacteria is 1 in 100,000. That's the same odds of the earth being hit by a half-mile-wide asteroid! Disinfectants eliminate all microbes on the spot before muta-tions have a chance to occur.

It has never been shown *outside* a laboratory that biocide products are linked to resistance to antibiotics. The few experiments showing resist-ance also show that its occurrence is too rare to be a health concern. In laboratory experiments, the favor-able conditions *encourage* the growth of resistant bacteria. Bacteria in the home grow much slower, and so the chances of mutating to a

they multiply they pass along their resistance genes to the next generation.

Some germicides leave behind a residue. Studies show that long-acting agents allow more time for the strongest bacteria to grow and develop resist-ance. These superbugs eject chemicals from their cells by a known mechanism called an efflux pump. They use this very same means to eject antibiotics. Chemical-resistant bacteria also resist the antibiotics ampicillin, tetracycline, chloramphenicol, and ciprofloxacin. A strain of *Staphylococcus aureus* has now been shown to resist both the antibiotic methicillin and the disinfec-tant benzalkonium chloride. This proves the antibiotic-chemical link.

Some situations do call for extra cleaning. For persons at risk of infection, after spilling raw meat juices, or even if the toilet overflows—the best things to use are 5.25 percent sodium hypochlo-rite bleach, 3 percent hydrogen per-oxide, or 70 percent ethyl- or isopropyl alcohol (rubbing alcohol). These are very effective and leave no residue.

Most bacteria are harmless. Elimi-nating harmless microbes along with the occasional pathogen goes against the "hygiene hypothesis." That is, a home that is too clean and disinfected hinders the normal functioning of the immune system. Exposure to a variety of microbes helps the immune system mature and produce antibodies, espe-cially true for the young. Doctors are seeing an increased incidence of asthma and allergies in children from homes that are too diligently cleaned.

In 2000 the *New England Journal of Medicine* showed that children from farms or owning a dog, or from large families, are all at lower risk for allergies resistant form are almost nonexistent. Antibiotic resistance is entirely the sad result of decades of overprescribed drugs.

Disinfectants are more important today than ever before. Global travel is common and increases the spread of disease between continents, threatening a pandemic outbreak. Also, food processing is becoming ever more centralized, increasing the chances for pathogens to contaminate meat and vegetables, fruits and juices. Finally, germs will spread easily as our urban populations grow. This phenomenon of urban spread of infection has been known since the Middle Ages. Antimicrobial products are increasingly essential.

In summary, current travel and food production encourage rapid spread of new disease. These are real threats. The media has sensationalized the public's fear of theoretical bacteria that are impervious to disinfectants while they rampage in our homes. They draw on terms such as resistant "superbugs." Scares over household cleaners are further fueled by academic researchers who run test-tube experiments rather than study real-life conditions. The more they talk about superbugs, the more attention they draw to themselves and their universities. Don't believe the hype.

than their counterparts from excessively clean homes. A steady exposure to a variety of benign microbes is good.

For generations, mothers have intuitively understood the hygiene hypothesis before it had a name. They practiced group vaccination by keeping all their children home when one came down with a communicable disease such as measles, mumps, or chicken pox. The resulting "family" immunity lasted almost a lifetime. Today, aggressive vaccination programs, disinfection, and hypervigilant cleaning have removed the opportunity for youngsters to develop natural immunity.

About 80 percent of infectious disease comes through person-to-person touch. For instance, a handshake followed immediately by a touch to the face transmits infection. Disinfectants are not used on hands so they do not affect this major route of germ transmission.

Good personal and home hygiene helps ward off infections. Proper hand washing, special care when sick with colds or flu, discarding used tissues, careful handling of foods, use of clean sponges and cloths, and well-maintained vacuums, are adequate for a germ-safe environment. Weigh the risks within your family and home setting. Then follow common sense before using strong biocides.

In summary, the media have sensationalized the idea of households crawling with a virulent mixture of viruses, bacteria, and mildew. Despite the publicity, which often comes from companies that sell biocides, there is no proof that people routinely catch germs from their laundry, doorknobs, the refrigerator handle, or the telephone. Don't believe the hype.

What's Inside the Can or Bottle?

The chemicals in disinfectant or sanitizer formulas that kill microbes are the **active ingredients.** All the other ingredients that suspend, dissolve, or propel the active ingredient are **inert ingredients.** Fragrances and colors are also inert ingredients.

The common antimicrobial ingredients you may see listed on a product are the following:

- Ethyl benzyl ammonium chloride
- Benzalkonium chloride
- Benzethonium chloride
- Sodium hypochlorite (bleach)
- Dimethyl benzyl ammonium saccharinate (or chloride)
- Various sulfonic acids
- Phenols
- Pine oil
- Alcohol, usually ethanol
- Triclosan

Antimicrobial active ingredients have different levels of effectiveness. For instance, sodium hypochlorite requires much less contact time to kill microbes, sometimes less than a minute, than most other chemicals. The first three names in the antimicrobial ingredients list belong to a group called quaternary ammonium compounds (nicknamed "quats"). Quats need more time than bleach to work and are generally less effective against Gram-negative bacteria. They are also sensitive to hard water. When cleaners with quats or phenols are diluted in hard water, the microbe-killing action may become weakened.

To complicate matters, microbes can be ranked in a hierarchy from the hardest to destroy to the easiest. Bacterial spores—*Clostridium* and *Bacillus*—are so resistant to damage that most

Water Hardness and Use of Quaternary Ammonium Biocides

Your city water-quality report gives you an estimate of the hardness of the water in your community. Hardness is caused by calcium and magnesium content. The hardness range is usually expressed as ppm (parts per million) of $CaCO_3$ (calcium carbonate). Other units used are mg/L, which is equal to ppm, or grains (1 grain = 17.1 ppm).

0–60 ppm	Soft
61–120 ppm	Moderately hard
121–180 ppm	Hard
Over 181 ppm	Very hard

Hardness affects only quats or pine oil products that must be diluted before using. The product label will give instructions for using the product in hard water. (One ppm is about two and one-half quarts of water in an Olympic-size swimming pool. One ppb, part per billion, is about an ounce of water in that same pool.)

disinfectants cannot destroy them. Only very strong products called sterilizers, or sporicides, work against bacterial spores.

The following infectious agents are ranked in order from the most difficult to kill to the easiest to kill:

Prions—The most indestructible of all infectious agents.
Bacterial spores—Species belonging to *Bacillus* and *Clostridium* form an almost indestructible spore coat.
Mycobacteria—Unique bacteria in structure and lifestyle.
Viruses without a lipid coat—These include hepatitis A,

rhinovirus (cold virus), polio, rotavirus, enteroviruses and Norwalk viruses, echovirus, adenovirus, and coxsackievirus.

Fungi—Includes *Candida* yeast and the molds *Trichophyton, Stachybotrys, Aspergillus, Penicillium, Fusarium,* and *Mucor.*

Bacteria that do not form spores—Important names are *Staphylococcus, Salmonella, E. coli, Streptococcus, Pseudomonas, Listeria, Campylobacter,* and *Enterococcus.*

Viruses with a lipid coat—Viruses coated with lipids, which are fatty substances, may be susceptible to certain chemicals that are good at infiltrating the fatty coat. These include influenza viruses (includes avian flu H5N1), AIDS virus (HIV), hepatitis B and C, herpes simplex (herpes) and herpes zoster (shingles), respiratory syncytial virus (RSV), hantavirus, varicella virus (chicken pox), coronavirus (includes SARS), and cytomegalovirus.

Some of today's most feared viruses are also the easiest to kill with chemical disinfectants. Viruses such as influenza, avian flu H5N1, SARS, and the AIDS virus are all susceptible to antimicrobial chemicals.

Household bleach is the WMD—weapon of microbial destruction. Bleach ruins virtually all microbial functions quickly (in seconds) even when diluted (according to manufacturers' instructions!). It kills two of microbiology's most robust members, bacterial spores and *Cryptosporidium* cysts, but for these tough microbes bleach needs long contact times, that is, hours rather than minutes. Despite bleach's reputation as a noxious chemical, in water a molecule of bleach degrades readily to a water molecule and sodium chloride, known better as table salt. Perhaps bleach's only disadvantages are (1) once diluted in water, its activity doesn't last long—bleach solutions should be made fresh daily—and (2) it is corrosive. Bleach doesn't clean things; it only disinfects.

Therefore, when using bleach you must first wipe or clean the dirty surface.

Antimicrobial Soaps

Antimicrobial soaps have received their share of the blame for causing chemical-resistant bacteria. Yet because they are soaps, they help block the spread of germs transferred from person-to-person.

Many people are justifiably confused by the terms "antimicrobial soaps" and "antimicrobials." Antimicrobial soaps include (a) hand or body soaps with an ingredient that prevents bacterial growth, (b) hand sanitizer soaps, and sometimes (c) antiseptics. Antimicrobial soaps are for the body and are not for cleaning things around the house. Antimicrobials are the disinfectants and sanitizers. They are *not* for use on the body.

Antimicrobial soaps contain either the ubiquitous triclosan (Table 5.1) or a related chemical called triclocarban. The packaging stresses that they are only for the skin by including claims that state,

Products that Contain Triclosan

Solid and liquid soaps	Cutting boards	Foot warmers
Daily face washes	Knives and slicers	Ice-cream scoops
Ear plugs	Acne creams	Toothpastes
Toothbrushes	Wipes and sponges	Dishwashing liquids
Deodorants	Shave gels	Sunburn sprays and creams
Skin care products	Makeup foundations	Lip colors and glosses
Burn treatments	Powders	Aprons
Mop heads	Furniture	Carpet cushions

"Helps eliminate germs on hands" or "For washing to decrease bacteria on the skin."

Do they work? Yes and no. Yes, they work like regular soap by breaking up oils on your skin to help wash away dirt. If the antimicrobial soap stays in contact with your skin long enough, it may kill some bacteria. No, they do not work any better than regular soaps because people rarely leave it on long enough to enjoy the extra benefit from the active ingredient.

Soaps that claim to kill germs are classified by the FDA as over-the-counter drugs. Understanding the terminology is far less important than understanding the need to use soaps. At the appropriate times and with proper usage, antimicrobial soap and regular soap are equally important in personal hygiene.

"Green" Household Cleaners

Green, or environmentally friendly, cleaners tout their safety for use around the home, and both the contents and the packaging are for-

Computer keyboards	Mouse pads	Towels
Paint and wall coverings	Handrails	Toys
Toy computers	Sandals, shoes, and socks	Shoe inserts
Air filters	Humidifiers	Flooring
Shopping cart handles	Pet dishes	Beverage pitchers

Table 5.1. Triclosan is used as an antimicrobial ingredient or coating.

mulated to decompose in the environment, that is, they are biodegradable.

Formulas not containing synthetic biocides are seeing renewed interest for two reasons. First, people are increasingly concerned about the persistence of chemicals in the outdoors and their fumes in the air indoors. Second, worries about microbes resistant to man-made biocides have prompted many consumers to turn to alternative cleaners.

The active components found in green cleaners sound pleasing to the ear: ingredients such as orange oil, lemon and lemongrass oils, tea tree and pine extracts, and surfactants (a type of detergent for breaking up dirt) made from coconut. If you read carefully, however, you'll also occasionally find amine oxide, propylene glycol ether, or sodium percarbonate.

Plant extract oils have been known for decades to inhibit bacteria. Therefore, spice oils and plant extracts are added to give green and some nongreen products their fragrance and also deliver a small amount of antimicrobial power.

Sodium bicarbonate (baking soda) and glycolic acid are added to some eco-friendly cleaners. Sodium bicarbonate brings a small amount of cleaning action, as well as a modest antimicrobial effect. Glycolic acid also acts against microbes and helps the other ingredients work better.

Green cleaners do not pass the EPA's strict requirements for claiming they kill microbes. Therefore, they may not have the words "germicide," "antimicrobial," "fungicide," "disinfects," or "sanitizes" on the label. The makers of green products are not permitted to list the names of microbes on the label. Most products do provide usage instructions and safety warnings. The irony becomes clear. Green products are purchased because they are viewed as safer than chemicals, for people, animals, and the environment. The EPA-registered antimicrobial chemicals are, however, the only ones that ensure they

Antimicrobial Essential Oils

Many industries use essential oils for adding flavor or fragrance to their products. Many of these oils also have antimicrobial properties.

Essential Oils Effective against Bacteria		Essential Oils Effective against Mold
Thyme	Lemon	Cinnamon
Marjoram	Oregano	Black pepper
Lavender	Laurel	Rosemary
Bay	Orange	Clove
Basil	Rosemary	Allspice
Cinnamon	Fennel	Sage
Nutmeg	Spearmint	Tarragon
Clove	Eucalyptus	Caraway

Because essential oils' activities are specific, each of these examples may be very effective against one species of bacteria or mold, but ineffective against another.

have been tested for both efficacy and safety, with data to back it up.

Many environmentally conscious people prefer to clean surfaces in their home with vinegar, baking soda, lemon juice, ammonia, or borax. Some people exclude ammonia and alcohol because they feel the pungent vapors are too dangerous to be considered people friendly. Undiluted vinegar, undiluted ammonia, and baking soda kill some bacteria, but they don't work as quickly or effectively as chemical disinfectants, especially bleach.

Think of green cleaners as roughly equivalent in effectiveness to registered germicides that aren't used properly. In either case, the product may eliminate a few thousand bacteria and viruses. In certain situations, a few thousand may be all you need to kill, but you have no way of knowing.

Families choose green products to meet their own requirements for safety, environmental responsibility, and effectiveness. You may be willing to accept a measure of uncertainty in microbe killing in return for knowing these formulas have been around a long time and are safe. If vinegar cannot kill 99.9 percent of bacteria within five minutes, but *can* destroy quite a few, is that good enough?

Finding an answer is difficult, as this chapter has shown.

Good Personal and Household Hygiene

(1) Wash hands frequently and thoroughly; (2) wash hands before you eat and after using a bathroom; (3) do not touch hands to mouth or eyes; (4) discard tissues after using; (5) avoid being near others who are visibly sick with a cold or flu, or people who don't cover their mouth and nose when sneezing/coughing; and (6) clean up around the house using *clean* sponges or *disposable* cloths.

These universal guidelines are recommended by the CDC, the Department of Health and Human Services (HHS), state health departments, and the Association for Professionals in Infection Control and Epidemiology (APIC).

Regardless of your choice in cleaners, your chances of getting sick are reduced if you return to the core principles of good hygiene.

How Bacterial Growth Affects Antimicrobial Action
When a single bacterial cell dives into a nutrient-rich liquid, it

doesn't immediately begin dividing and growing. The single cell goes through a period of sluggish growth during which it makes enzymes and builds other components in preparation for the coming burst of activity. This **lag period** may last from hours to days. Once the cell assembles all its systems for replication, it begins dividing and living at the doubling rate characteristic for its species. The rate may be as short as twenty to thirty minutes. Short doubling times cause the size of the bacterial population to rapidly expand. (Refer to Figure 2.3.) This period of extremely fast growth is the exponential phase or logarithmic phase, often shortened to **log phase.**

Microbiologists are the rare scientists who daily use the logarithm, which is a mathematical operation for converting extremely large numbers to versions that are easier to manage. Without logarithms, subtracting, dividing, or any other calculation performed on these large numbers would be cumbersome to the point of impossible. Luckily, most hand calculators do the fancy arithmetic. An example of log conversions is shown here for the mathematically inclined:

$2 \times 10 = 20$
$2 \times 10^3 = 2000 = \log 3.3$
$2 \times 10^6 = 2{,}000{,}000 = 2 \text{ million} = \log 6.3$
$2 \times 10^{10} = 20{,}000{,}000{,}000 = 20 \text{ billion} = \log 10.3$

There are a few places other than the test tube where bacteria could approach logarithmic growth. A cup of milk not refrigerated, a still backyard pond on a hot summer day, or an open wound bathed in blood and grime are examples of sites where bacterial populations may increase very rapidly. But in average circumstances, bacterial growth tends to be slow around the home or on the body. This slow rate of growth affects disinfection.

Rapidly dividing cells are more susceptible to damage from

drugs, biocides, and environmental extremes than slow-growing or dormant cells. The cells use most of their energy resources during logarithmic growth to build new exterior walls, enzymes, and chromosome apparatus. Disinfection studies on log-phase and non-log-phase bacteria suggest that antimicrobial compounds work best when cells are rapidly dividing, that is, the log phase.

In a household environment microbial multiplication is not fast unless there is a fortuitous new supply of nutrients, that is, a glass of milk spilled on a carpet. The microbes getting by on limited provisions suddenly receive a bounty and burst into high activity for a short period of time. Usually, however, household conditions are not superb. Temperature and moisture levels may be disagreeable. Nutrients may be limited. Microbes have to get by with the conditions at hand.

In the outdoor or indoor environments, microbes slow their cellular activity and divide into new cells only if enough nutrients are available to sustain them. Eventually, new cells are created at the same rate as old ones die. This is the **stationary phase.** Many microbes can better withstand chemical biocides in stationary phase.

Disinfectants are formulated using optimal test-tube conditions in a laboratory. Products are tested on bacteria during their logarithmic growth. Therefore, the antimicrobial product is given every advantage to work at peak performance. The products are rarely tested on the mix of microbes clinging to your cutting board. Perhaps those words claiming "kills 99.9 percent of germs" or "disinfects against . . ." are overestimating the horsepower inside the bottle.

The world doesn't fill up with dead bacteria any more than it fills up with corpses and carcasses. Decomposition keeps our planet livable—decomposition carried out by vibrant communities of microbes. Once stationary cells reach a point where nutrients are gone and end products build to poisonous levels, they enter **death phase**. Here they stop functioning and quickly break apart. Cell contents dissolve into surrounding fluids, and the pieces that don't dissolve litter the microscopic landscape. In fact, dead microbes represent some of the household dirt that prevents disinfectants from working at their best.

In the laboratory, bacteria's "circle of life" takes place in a few days. A clear and sterile broth solution is inoculated. The tube is put into an incubator. The broth in the tube may remain clear for several hours or for a day or two, depending on the type of bacteria. Quickly though, the liquid begins turning murky—logarithmic growth. The broth gets cloudy, dark, and gives off aromas you will not soon forget. But the activity slows over time, and the culture begins to die. If left in the incubator long enough, the broth turns almost clear again. Sometimes, all that is left is slight cloudiness and a faint odor.

You don't have to ponder the philosophy of life to understand basic disinfection. Disinfectants and sanitizers are invented under laboratory conditions that are most favorable for showing effectiveness. If the directions state the formula should remain on a surface for five minutes, you may assume that it will not kill all the desired microbes if you leave it for less than five minutes. And because your home rarely represents perfect disinfecting conditions—there may be dirt, unknown microbes, or stationary phase cells—you may also assume that letting the product remain on the surface a few extra minutes before you rinse it off is a smart thing to do.

Antiseptics

Biocides, virucides, sanitizers, and mildewcides, oh my! Thank goodness for simple things like antiseptics, which cause no confusion at all. Antiseptics are biocides that kill microbes. Therefore they are antimicrobial. Nevertheless, they are not germicides. Antiseptics are used on the skin and not inanimate objects, so the FDA classifies them as drugs rather than germicides. Drugs control disease conditions. But how is an alcohol swipe prior to an injection considered disease control? It's because the FDA further defines "drug" as any substance for use in diagnosis, cure, mitigation, treatment, or prevention of disease or for affecting body function. Antiseptics *mitigate* the spread of germs. For that reason, an antiseptic is a drug. Some antiseptics are also considered antimicrobial soaps, a category that does not include hand washes. Hand washes are, however, a subcategory of antiseptics. And the commonly used antiseptics include alcohol, hydrogen peroxide, iodine tincture, and the quat benzalkonium chloride, which you will easily remember are also disinfectants! Simple.

The term "antiseptic" was introduced more than 250 years ago and it should come as no surprise that its definition has suffered through about 200 iterations. Manufacturers are not perplexed; they put "antiseptic" on everything from sunblock to toilet seats! The FDA has chosen not to applaud this ingenuity and has now drawn up a detailed list of categories for various antiseptics (Table 5.2).

Currently Available Antiseptic Products

Product	Target Consumers	Use(s)	Where Used
Healthcare antiseptics			
Hospital handwash	Health care professionals, patients	Reduce skin bacteria prior to surgery or patient care	Hospitals, outpatient clinics, medical offices, nursing homes
Pre-op skin preparation			
Surgical hand scrub			
Hand sanitizer			
Antiseptic hand wash	General population	Reduce bacteria on hands	Homes, day care centers

Some active ingredients are formulated into both antiseptics for the skin and cleaning products for the home. An example is benzalkonium chloride found in antiseptic creams and in products that sanitize your kitchen. This *never* means that you may use the products interchangeably. Every antiseptic or disinfectant label will unfailingly tell you where the product is to be used.

The Truth about Alcohol

In Shakespeare's *King Henry the Fifth,* Henry lamented, "I would give all my fame for a pot of ale and safety," but any microbiologist prefers a trusty bottle of 70 percent ethanol at her side. Rubbing alcohol (70 percent isopropyl or isopropanol) works as well as ethyl alcohol (ethanol) and is easy to find in any grocery or drug store.

If 70 percent alcohol is good at killing bacteria and fungi, wouldn't

Consumer antiseptics			
Product	**Target Consumers**	**Use(s)**	**Where Used**
Antiseptic body wash		Reduce body odor	
Hand sanitizer		Prevent infection	
Food handler antiseptics			
Handwash	Commercial food handlers	Reduce risk of foodborne disease	Restaurants, processing plants, other commercial food establishments
Hand sanitizer			

Table 5.2. The Categories of antiseptic products as recommended by the FDA Center for Drug Evaluation and Research.

95 percent be better? Alcohol works quickly and leaves no residue after drying. But it dries quickly. Ninety-five percent alcohol evaporates so fast there is not enough time to destroy the microbe's proteins and fats. Diluting it with water slows evaporation, and this prolongs the contact time. Too much water, however, takes away alcohol's microbe-killing power. The effective concentration range is between 50 percent and 85 percent, and most labs and clinics have adopted 70 percent as the standard concentration for killings microbes with alcohol.

How Antimicrobial Chemicals Work

Antimicrobial compounds kill microbes by attacking normal functions needed for growing and multiplying. Cell membranes, cell walls, enzymes, and DNA-replicating mechanisms are some of the targets of germicides. Details of the modes of action for some antimicrobial compounds are not fully known.

Antimicrobial Chemical	How It Works	Works Best Against
Sodium hypochlorite (bleach)	Interferes with oxygen use, disrupts proteins, destroys membranes	Bacteria, viruses, protozoa, spores
Alcohol, 70 percent	Denatures proteins (breaks bonds that hold proteins in their normal structure), dissolves membrane fats	Bacteria, fungi, some viruses
Ammonia	Releases hydroxyl ion that interferes with cell activities, possibly nutrient uptake	Bacteria, viruses, fungi
Quaternary ammonium compounds ("quats")	Binds to membrane to disrupt function and release cell contents	Bacteria, some fungi and viruses
Metals (silver, copper, zinc)	Reacts with enzymes to poison the cell	Algae, fungi, some bacteria

150

When the nurse swabs your arm before jabbing you with a needle, the alcohol removes microbes on the skin. The bacteria and fungi are carried away with the swab, leaving a sterile patch of skin, right? No, because of transient and resident flora.

Resident flora are the native bacteria, yeasts, and other fungi that call your body home. They live in skin crevices or beneath the skin's uppermost layers. Not all of them are dislodged with alcohol or even with a vigorous scrub in the shower or a hand washing. **Transients** are

Antimicrobial Chemical	How It Works	Works Best Against
Acids (vinegar, lemon juice, etc.)	Changes acidity of cell's surroundings to disrupt normal functions	Bacteria, some viruses
Pine oil	In combination with other biocides disrupts enzymes and membranes	Bacteria, some viruses
Sodium bicarbonate (baking soda)	Interacts with normal acid-base balance	Some bacteria
Hydrogen peroxide	Strong oxidizing agent, it produces unstable forms of oxygen that destroy DNA, membranes, and other cell components	Bacteria, viruses, algae, some activity against fungi, yeast
Iodine	Interferes with amino acid structure, thus destroying protein function	Bacteria, viruses, protozoa, some activity against fungi and spores
Triclosan	Blocks enzyme used in building fatty acids	Bacteria

those you pick up, slough off with dirt and skin flakes, and pick up again a few minutes later. Swabbing is effective in removing transients. Years of use seem to confirm that alcohol swabbing prevents infection.

Alcohol hand washes are also effective and have become popular for managing germs during air travel, car trips, camping, or other places where water isn't available.

Here is the way to use alcohol hand washes:

1. Squirt the correct amount into the palm.
2. Rub hands together vigorously and make sure the alcohol comes in contact with all parts.
3. Be sure to include the fingertips and areas between the fingers.
4. Continue rubbing until the hand wash evaporates and the hands are dry (usually fifteen to twenty seconds).

Summary

Germicides kill microbes that may threaten the health of certain at-risk populations. They are useful in the home for disinfecting or sanitizing places where there are known to be high levels of microbes. Microbiologists disagree on how much cleaning and disinfection you really need in your life. Some feel that cleaning and disinfection must be held to a minimum; others believe a spotless house is the only way to go. In addition, scientists and biocide makers often debate the danger, if any, from resistant bacteria. As with any other product, before buying, consider the facts and your personal needs. Common sense and good judgment trumps almost any pitch you'll see on television.

6

Infections and Disease

There is no little enemy.

—Benjamin Franklin

Disease travels best through crowds. Communities of squirrels, calves, humans, trout, grape vines or oak trees, and every other living thing are susceptible to infection when individual members are close together.

Our urban communities have throughout history suffered higher rates of infection from microbes than did their rural counterparts. From ancient times to today, infectious diseases have often altered the paths of civilization.

The smallpox virus appeared in 1200 B.C. Egypt and continued to curse society for centuries. In the eighteenth century, five reigning European monarchs were killed by smallpox. Its eradication two centuries later was one of history's few worldwide cooperative efforts among governments and cultures. The devastation rendered by the great plagues of the Middle Ages are lessons in biology and history. The bubonic plague, or Black Death, killed 100 million people in the sixth century, 25 million in the fourteenth century, at times killing one in four people. The loss of life was so great that historians believe it slowed the pace of scientific and technical advances in Europe in the following centuries. In the present, could there be any question of the effects the AIDS epidemic has had on the alliance between government policy and health care doctrine? These are the briefest of examples of the extensive societal changes related to a microscopic speck.

Before an outbreak occurs in a population, a single virulent microbe must establish itself within a subpopulation. **Virulence** is a microbe's capacity to infect and cause disease, and the subpopulation are those people who, through circumstance or health, are vulnerable to infection by the microbe. These people will harbor the microbe and will likely contribute to the initial spread of disease. The body's immune system is one of our primary defenses against infection. Its function, plus a variety of other physical conditions within the body, defines a person's **susceptibility** to infection. Every person has some level of susceptibility that will change throughout a lifetime. Behavior, too, plays a role in giving pathogens the **opportunity** to spread throughout a community. There is little you can do to affect a pathogen's virulence. You can, however, influence your susceptibility and the opportunities you give to microbes that are poised to invade your body.

The Mechanics of Infection

Infection is the invasion and growth of microbes at a specific, localized area on or in the body. Some infections are mild and treated without a doctor's help. Others progress beyond their localized point of origin and spread in the body, thereby causing a specific **infectious disease. Diseases** are major events in which all or part of the body cannot function normally. The definition of "disease" is sometimes refined to include any event in the body that causes impaired function. By this broad standard, even an infection would qualify as a disease.

Diseases are categorized in a number of ways, and the classifications change over time with new information on the cause and progress of disease. Diseases can be classified according to the organ or tissue affected. For instance, Parkinson's disease is a *neurological* disease because it affects the brain. Alternatively, diseases are categorized by the agent or activity that serves as the primary cause.

Types of Diseases Classified by Their Causes

Infectious disease—Invasion of tissue by microbes, for example, influenza, AIDS, encephalitis

Inherited or congenital—Passed from parents to offspring, for example, Down syndrome

Dietary—Underfeeding, nutrient deficiencies, overfeeding, for example, pernicious anemia in vitamin B12 deficiency, obesity

Industrial or occupational—Initiated by specific behaviors, for example, repetitive motion injury, black lung disease

Environmental—Acquired from nonliving agents in air, water, foods, and so on, for example, lung cancer, melanoma, lead poisoning

Behavioral—Caused by habits or lifestyle, for example, lung cancer from smoking, cirrhotic liver from alcohol consumption

Mental illness—Neurological diseases with behavioral syndromes or anomalies, for example, schizophrenia, phobias

Degenerative diseases—Degeneration of tissue over time associated with combined factors of aging, nutrition, exercise, environment, or infection, for example, multiple sclerosis, Parkinson's disease

Worldwide, infectious disease ranks second in this list of major causes of mortalities: (1) cardiovascular diseases, (2) infectious diseases, (3) cancers, (4) injuries, and (5) respiratory diseases. Within specific pathologies, infectious diseases are the third and AIDS is the fourth leading cause of death after heart disease and stroke (cardiovascular). Within infectious diseases alone, AIDS, diarrheal diseases, tuberculosis, and malaria claim the most lives.

Diagnosis and medical care in the United States is superior overall compared with other regions in the world. The CDC ranks infectious disease as the seventh leading cause of death in the

United States. Of these, influenza and pneumonia cause the most deaths.

Entry and Infection
Port of Entry

The likelihood that a disease-causing microbe will successfully infect a host depends on (a) its ability to enter or adhere to host cells, and (b) the number of pathogens that assault the host. The factors relating to microbial attachment to and entry into human cells is a vast area of scientific study and draws on immunology, serology, and enzymology.

The body's mucous membrane linings are the major route of entry for bacteria and viruses when invading a host. (Mucus is a noun describing the thick material secreted by certain membranes and glands. Mucous is an adjective. For example, *mucous* membranes secrete *mucus*.) Among several factors that make mucous membranes a desirable entry point is the ease with which microbes can penetrate them.

Mucous membranes line the respiratory, digestive, and urogenital tracts, as well as much of the infrastructure supporting the eyes. The respiratory tract and gastrointestinal tract are the most common entry ports. Most pathogens have a specific mucous route they prefer: *Streptococcus pneumoniae* prefers the respiratory tract, *Salmonella* uses the gastrointestinal tract, and herpes virus enters through the cells lining the urogenital tract. By comparison, the tetanus microbe *Clostridium tetani* infects by attacking skin cells.

Certain microbes use more than one port of entry. One reason why anthrax is so feared is because the microbe is both deadly and able to infect through multiple routes. A common route is through a cut or skin abrasion. The bacteria may penetrate through deeper skin layers, but will rarely enter the bloodstream and remain a **localized infection**. Inhaling anthrax spores is less common than skin contact, but it is far more deadly because the pathogen enters the bloodstream. Inhalation anthrax kills close to 100 percent of those who contract it. A third rare but dangerous (mortality rate 50 percent), form is gastrointestinal anthrax, in which the microbe infiltrates the lining of the digestive tract, causing severe lesions.

Infecting the Host

Pathogens all have in common a process for attaching to, then getting inside, their targeted tissue or cell type. Adherence to a host cell involves a connection between the host cell's surface and molecules on the outside of the pathogen. This binding between host and pathogen is the first crucial step in infection.

Once inside the body, a pathogen must avoid the body's defense systems. Some disease-causing microbes have evolved to possess a few defenses of their own. Some form spores and others develop cell walls that are difficult to destroy. Other microbes secrete enzymes that digest host tissue to help the microbe gain entry into organs.

If a pathogen evades the immune responses, it may destroy an entire organ or metabolic system. (A metabolic system is a group of organs that serve to provide the body with a function required for life. For instance, the digestive system is a metabolic system.) The *Streptococcus* that causes necrotizing fasciitis (the flesh-eating bacteria) lives in and chews its way along the skin. It doesn't really chew, of course. It excretes an enzyme that breaks down skin for nutrients. Other bacteria gather certain nutrients such as iron from the blood or from organs. Eventually the body falls into a deficiency condition that leads to additional complications.

Some bacteria and viruses damage the body indirectly by decimating immunity. This leaves the person, who is then considered immunocompromised, vulnerable to other microbes and their infections, called a **secondary infection.**

In general, viruses and the tuberculosis bacterium penetrate human cells to cause disease. Other bacteria tend to adhere to the outside of tissues and organs.

Toxins

Toxins, rather than whole microbial cells, do the major damage in at least half of all bacterial diseases. A toxin is a poison made by a microbe. Sometimes bacteria are no longer present in the body, but their toxin remains and it circulates with the blood. These toxins excreted from the pathogen are exotoxins. *Clostridium botulinum* toxin works this way. The foodborne pathogen may be killed during cooking, but the toxin remains active and causes deadly botulism poisoning after being ingested.

In some Gram-negative bacteria, the toxin remains attached to the bacterial cell wall (an endotoxin) and there may be no symptoms until the cell dies and breaks apart (lysis) and the toxin is released. When antibiotics are prescribed for endotoxin-producing bacteria, you may feel worse shortly after taking the drug. This is actually a sign that the antibiotic is doing its job. The symptoms become more severe for a time as the newly released toxin circulates. Examples of endotoxin-associated diseases are typhoid fever caused by *Salmonella typhi* and meningitis caused by *Neisseria meningitides.*

Neurotoxins are those that attack the nervous system, disrupting nerve-to-nerve communication or that from nerves to organs. Symptoms include disorientation, involuntary muscle contraction, dizziness, memory loss, and other characteristics of nerve damage.

Infectious Dose

The higher the number of pathogens, the greater chance a person has of being infected. The number differs from pathogen to pathogen. Infectious dose is the approximate amount of a microbe that will cause its specific disease.

If the body is exposed to an infectious microbe at numbers much lower than the infectious dose, the immune system will likely destroy the few interlopers. At higher doses, pathogen adherence, entry, and disease are more likely to take place.

Some Infectious Doses

MICROORGANISM	DOSE
Tuberculosis bacterium	3 cells
Hepatitis A virus	Less than 10 viruses
Norwalk virus	10 viruses
Giardia	1 to 10 cysts
Cryptosporidium	10 to 100 cysts
Salmonella species	About a million cells, but may be much lower (100–1,000) in fatty foods like chocolate or cheese
Salmonella typhi (typhoid)	10 to 100 cells
Campylobacter	About 500

Table 6.1. Infectious dose varies by pathogen.

Infectious dose is variable among different people because of their personal health and susceptibility. A preexisting disease makes you more vulnerable and exposure to fewer pathogens may cause infection. If you are at peak health, it might take a slightly higher dose of pathogens to make you sick. Considering the Five-Second Rule, you should therefore mull over your health history before biting into the cookie recovered from the floor.

Disease Transmission
The Many Ways to Transmit a Germ
Bacteria do not fly. Viruses can't swim. A yeast cell cannot leap the tallest buildings. Yet microbes manage to make their way from place to place, person to person. Many are able to endure stopovers in barren wastelands. Knowing a bit about how germs get around on their own or with unsuspecting help is valuable in avoiding infection.

During flu season, outbreaks in schools, nursing homes, dormitories, and athletic teams seem to make the news each year. It's not difficult to pinpoint the site of the outbreak in static locations where people are in proximity. Places where people are in transition, such as subways and buses, also spread infection, but it's harder to trace an outbreak to these sites and ever-changing crowds.

During the Middle Ages urbanites enduring the plague seemed to understand the risks of nearness. As plague victims died in homes and in the streets, the bodies were carried out of town to be burned. Handling the infected seemed unwise. The villagers therefore devised a means of transporting the bodies by spearing them with poles at least ten feet long and carrying the kebab to the disposal site, presumably muttering along the way, "I wouldn't touch him with a ten-foot pole."

Stopping your hand-to-face habits foils infectious microbes. Adults touch their eyes, nose, and mouth an average of three times every five minutes. Children do it more than twice that number. Estimates of the total daily hand-to-face touches range from twenty to hundreds. Pathogens lying in wait on solid surfaces or on your hand—remember how truly unsanitary is the handshake—are picked up and deposited at a pathogen's favorite port of entry, mucous membranes.

Microbes expelled by coughs and sneezes travel up to three feet through the air in moisture droplets. You may catch a cold by standing on the receiving end of someone's sneeze, but it is more likely that the virus will land on a surface around the house or office, and then you'll pick it up on your hands. Cold viruses and flu virus can remain infectious for up to three days on an inanimate surface, and flu remains dangerous for up to an hour in aerosol droplets. Fecal microbes, *Staphylococcus,* hepatitis A, and rotavirus also persist for hours to days on any surface that has not been thoroughly cleaned and disinfected or sanitized.

The Best Defenses against Spreading Infection

The best defense against infection is to follow the principles of good hygiene (in Chapter 5). It is wise, too, to be aware of the varied modes of travel pathogens use to get from there to here.

The public health profession has expanded the picture of disease transmission quite a bit from the days when people believed sickness came as punishment for consorting with the devil. **Vertical transmission** occurs between a mother and her developing fetus. **Horizontal transmission** is the passing of infection from one person to another. A subcategory of horizontal is **direct transmission**, which is body-to-body contact, such as sexually transmitted diseases.

Most people understand the risks of infection through person-to-person contact. Often overlooked is **vehicle transmission**, the spread of germs between people or animals by way of an intermediary. When transmission occurs through an inanimate (nonliving) object, it's called **indirect transmission** (also indirect contact transmission). The term for the inanimate object that serves as the microbe's temporary stopover is **fomite** (Figure 6.1). Think toilet flusher, remote control, hotel phone, refrigerator door handle, and so on.

The Five-Second Rule depends on indirect transmission. A family member may step in dog waste in the yard, then track a small amount onto the kitchen floor. A short time later, you drop a cookie to the same floor. As you pick up the cookie and consider taking a bite, remember indirect transmission.

Reservoirs and Vectors

Air, water, and hands all serve as vehicles. Insects, too, are vehicles, but when infection is transmitted by an arthropod, it is referred to as **vector transmission** and the arthropod is the vector, or intermediate carrier, of the pathogen. Vectors get pathogens from a **reservoir**, a continual source of a specific infectious agent.

Those pole-toting citizens during the plagues unwittingly

Body Art

Tattoos and body piercing date as far back as 3300 B.C. Today, about 25 percent of Americans have tattoos. Though a larger percent (36 percent) of those born after 1975 than before have tattoos or piercings, the popularity is spreading to older ages. Grandparents and professional career people are no longer outside the body art boundary. Tattoos are slightly more popular than piercings, while almost three times more women than men have piercings.

Attention to safety and cleanliness is improving in body art emporiums. Still dermatologists point out the hazards inherent in any activity that damages the protective barrier of the body's skin. Worrisome, too, is an advancing incidence of MRSA from unlicensed tattoo parlors. Some states require parlors to be licensed, but patrons still go to unlicensed artists.

Tattoos

Tattooing involves depositing pigments from oils or synthetic dyes into the skin at a depth of 1 to 2 millimeters, about three-sixteenths of an inch. Native flora and infectious pathogens enter the newly created wound during the process. Immediately after tattooing local infections are often experienced, characterized by redness, swelling, and pain. On occasion, bacteremia (bacteria in the bloodstream) ensues. Infections are typically from *Staphylococcus aureus*, *Pseudomonas aeruginosa*, and *Streptococcus pyogenes*. *Clostridium* infection (tetanus) is associated with sticking and probing by sharp metals, but fortunately, most people are vaccinated against tetanus.

The artist is often the source for hepatitis B and C infections. A University of Texas Southwestern Medical Center study found people with tattoos were six times more likely to have hepatitis C than people without them. For this reason the American Association of Blood Banks prohibits blood donations until at least one year after any tattoo procedure.

Body Piercing

The incidence of infections localized at pierced sites ranges from 10 to 30 percent. The following sites are generally associated with certain infections, usually from native flora: ear (*Staphylococcus aureus*, streptococci, *Pseudomonas* bacteria); nose (*S. aureus*); tongue (oral bacteria, *S. aureus*, papillomavirus); nipple (staphylococci, streptococci); navel (*S. aureus*); and genitals (papillomavirus). Though they are not native microbes, hepatitis B and C are concerns, and the American Academy of Dermatology (AAD) cites tuberculosis and tetanus as additional risks in piercing and tattoos, as well as yeast-infected abscesses. Cartilage piercing, such as that in the upper ear, is slow to heal if infected because its low blood supply prolongs the body's healing process.

Overall, public health experts are divided on the actual dangers of body art. Infections are not always reported, so hygiene-challenged parlors are hard to identify. Body art parlors in the United States are not government regulated, and the instruments and dyes are not reviewed by the FDA for sterility and safety.

The AAD offers these tips for reducing infection risks with body art:

TATTOOS	Uncover the dressing after twenty-four hours to expose the wound to the air.
	The wound may be cleansed with mild soap and water.
	Do not use **alcohol-based** lotions to moisturize the area.
	Stop using antibiotic creams if they cause irritation.
	Protect the tattoo from direct sunlight until fully healed.
PIERCINGS	Swelling after tongue piercing can be reduced by sucking on ice cubes, or rinsing with saltwater or mild mouthwash.
	Cleanse lip and navel piercings with soap and water.
	Cleanse genital piercings with warm saltwater.
	Avoid wearing tight clothes over piercings until they heal.
	Ensure the metals used are nickel-free surgical steel, titanium, or niobium. Inferior metals corrode and may cause infection.
BOTH	Note cleanliness of the parlor and ask the artist questions on how equipment is cleaned and stored.
	Make sure the artist wears mask and surgical gloves, uses new needles for each new customer, and uses equipment and needles that have been sterilized for each customer. (Look for equipment and needles that are wrapped like surgical or dental instruments.)
	Make sure the artist swabs your skin with alcohol before starting the process.

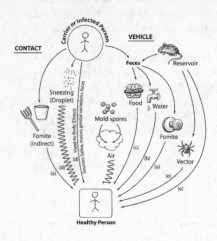

Figure 6.1. Disease Transmission. There are several points where germ transmission can be blocked. (a) Proper cleaning, disinfecting. (b) Proper water treatment. (c) Cooking and safe handling of foods. (d) Avoid hand-to-face touch. (e) Caution with animals and insects. In *all* instances, handwashing is important! *Illustrator, Peter Gaede.*

demonstrated the vector-reservoir relationship. Today, as in the Middle Ages, the microbe *Yersinia pestis* uses rodents as its reservoir. Fleas feeding on them ingest the *Yersinia* bacteria, and then carry it to others with each additional bite. Not only does the flea begin the disease's journey toward humans, but it also serves to maintain the concentration of *Yersinia* in its rodent reservoir.

A nip from a rat, or more likely a bite from a flea vector, is the point at which the infectious disease makes its animal-to-human jump. Plague may reemerge as a concern as communities encroach into undeveloped areas and live near the pathogen's reservoir. Today's domestic cats are additionally implicated in modern *Yersinia* outbreaks if they consume wild rodents that are infected.

Y. pestis contributes its own clever ways to ensure plague persists. Interestingly, it makes the fleas sick by blocking their digestive action. The starving fleas' hunger cannot be satisfied so they seek more and more blood and thereby increase the size of the rodent reservoir. Once inside humans, the bacteria live inside phagocytes, the very cells the immune system sends into the bloodstream to seek and destroy invading microbes.

The plague scenario illustrates vehicle transmission and vectors. It also demonstrates the macabre influence disease has on our society. Plague lesions are rosy pink before turning blackish-blue, of musical importance as well as medical. The nursery rhyme "Ring Around the Rosy" is thought to have originated in the Middle Ages to describe the lesions, and youngsters gleefully sang it as the Black Death spread across Europe.

Virulence and Susceptibility

When the body's immune system is weakened, pathogens have an easier time starting an infection. A damaged immune system gives even normal flora the opportunity to become dangerous. The AIDS epidemic gave new meaning to **opportunistic pathogens**, normal flora that turned deadly because a damaged immune system could not defend against them. In other words, normally harmless flora had been given the *opportunity* to cause disease. Patients in the early days of the epidemic had visited hospitals and doctors' offices with a variety of obscure ailments caused by "new" microbes. Doctors struggled to identify these opportunistic pathogens even as hundreds of the infected continued to fall victim to them. (Some of the opportunistic infections in the early part of the epidemic were *Pneumocystis pneumoniae,* retinitis from cytomegalovirus, viral encephalopathy, protracted diarrhea from *Cryptosporidium,* and *Cryptococcus meningitis.*)

All households will at some time have a member in a high-risk health condition or someone with a cold or the flu. Good personal hygiene and housekeeping are important steps to reduce the spread of pathogens. Being mindful of transmission methods is just as essential.

You cannot blame your cold on urbanization alone. The United States is a service-based economy. Office buildings and mass transit systems are crowded. Airport terminals and some schools have become overcrowded (as have prisons). Health specialists are citing the trend toward office cubicles as a factor in spreading infection. Add to the mix an aging population and an older workforce, some with compromised immunity. The result is a population in which heightened susceptibility and increased pathogen opportunity come together.

Hospital-Associated Infections

Hospital inpatients and outpatients contract infections at a higher rate than the general population. About 2 million people per year are infected with a microbe they got in a hospital, a **nosocomial** infection. Nursing homes and outpatient clinics are implicated along with hospitals.

Nosocomial infection rates have risen dramatically over the last twenty years. More alarming is the fact that the major known microbial causes are *all* antibiotic resistant. A few of these bacteria are now resistant to third-generation antibiotics.

Nosocomial infections tend to be opportunistic. In health care settings, many people are concentrated in a confined area and hand-to-body touching is frequent. The proportion of sick patients is high, and preexisting disease has weakened their immune systems. Some therapies add to the problem by temporarily weakening the immune system: organ transplantation and chemotherapy for cancer. In hospitals, damage to one or more physical barriers is common. Broken skin or mucous membranes from trauma, injections, surgery, or burns provide an entry port for infection. Urinary catheters, intravenous lines, and ventilators additionally increase risk. Extra care to reduce infection is warranted when using such devices.

A hospital can act as its own unique environment for harboring antibiotic-resistant microbes. Specific problem bacteria may even be localized to wards within a hospital and contain a strain not found anywhere else in the building. Overall, urinary tract infections, skin infections, and infections at surgical sites predominate, followed by respiratory infections and those caused by catherizations.

Forty to fifty years ago antibiotic-treatable *Staphylococcus* was the main, and only, known microbe causing nosocomial infections. Today, antibiotic-resistant *Staph* predominates and is joined by several other species. Among the major hospital-associated infections, *Klebsiella pneumoniae* has increased the fastest within the last five years.

Immunity

Immunity is the body's system for defending itself against non-native microbes. Your susceptibility to infection is determined by the strength of your immune system. Perhaps because of the critical role

Major Hospital-Related Infections:

Methicillin-resistant, coagulase-negative *Staphylococcus*

Methicillin-resistant *Staphylococcus aureus* (MRSA)

Ciprofloxacin/Ofloxacin-resistant *Pseudomonas aeruginosa*

Levofloxacin-resistant *P. aeruginosa*

Cef 3-resistant *Enterobacter* (Cef 3 = third generation Cephalosporin)

Penicillin-resistant *pneumococci*

Vancomycin-resistant *Enterococcus*

Cef 3-resistant *Klebsiella pneumoniae*

Imipenem-resistant *P. aeruginosa*

Quinolone-resistant *E. coli*

The CDC's National Nosocomial Infection Surveillance system publishes trends and its Web site provides guidelines for both patients and health care workers in the areas of hand hygiene, invasive devices, catheters, and surgical sites.

The hospital mimics the general population in at least one important way: touching is the cause of at least 80 percent of the infections. Visitors, patients, and non-nursing staff must be as aware of germ transmission as are nurses and doctors. Even among hospital staff, it has been shown that less than 30 percent of all facility workers wash their hands before being near patients. Despite more hygiene specialists hired by hospitals, the incidence of nosocomial infections continues to rise.

Florence Nightingale once offered clearheaded advice. "It may seem a strange principle to enunciate as the very first requirement in a Hospital that it should do the sick no harm."

the immune system plays in protecting the existence of species, it has evolved a number of components, some complementary and some that back up first-line defenses.

Innate Immunity

A disinfectant protects you from a pathogen only until another one comes on the scene. That may happen an hour after using the disinfectant. It may happen within a minute. Our immune system, by contrast, provides protection against a multitude of pathogens for years or for a lifetime.

The purpose of the immune system is to be on guard for any foreign particle that may invade the outer barrier of skin or mucosal surfaces. In broad strokes, a bit of pollen or a cold virus is the same to our body's way of thinking—they are invaders and they must be destroyed. (True also in organ transplantation, where immense efforts are made in fooling the immune system into identifying "foreign" as "self.")

There are two parts to the immune system. One is the innate system. It develops at birth or very soon after. The second is acquired immunity, that which develops in response to a foreign entity. Both systems provide the body with its second line of defense if skin and mucosal membranes are compromised.

Imagine a cluster of *Staphylococcus* cells on your arm. You get a cut, deep enough to bleed. The cutting action pushes the bacteria through the skin layers, and they are swept up by the flow of blood. If the body doesn't react, the small band of microbial invaders will multiply. The body provides them a wonderful supply of meals and garbage pickup, courtesy of the bloodstream, as well as a warm, steady climate. Invasion of the blood system by microbes is called **septicemia**. It is the immune system's job to make certain septicemia never happens. Should the microbes multiply in the blood, spread, and induce an inflammatory response, this serious condition is called **sepsis**.

Specialized white blood cells (leukocytes) detect the bacteria in the bloodstream. The leukocytes alert others that are more equipped to destroy bacteria, and the *Staphylococcus* cells are devoured by phagocytosis (or phagocitized) within minutes. If a few staphylococci remain behind in the area of the cut, other white cells arrive and set up an inflammatory response. Histamines are released, causing blood vessels to dilate and allowing more leukocytes to rush toward the invaders. All the while, the body has set up a fibrous wall around the injury site to prevent the invading bacteria from escaping. Cornered and defenseless, the *Staphylococcus* cells are destroyed.

Some *Staphylococcus* strains produce coagulase, an enzyme that coagulates blood shortly after an injury occurs. The thickened blood forms a barricade around the bacteria and provides them with a temporary fortress against detection by blood leukocytes. The delay gives the *Staphylococcus* extra time to multiply, increasing their chances to advance the infection.

What if one or two bacteria manage to evade the attacks by leukocytes in the bloodstream? In this situation, another type of white blood cell in the tubes and nodes that make up the lymphatic system is employed. The lymph system removes any debris in blood as it slowly filters through its channels. Bacteria flowing through the ducts are seized, and then the lymphocytes kill them.

In addition to the cellular immunity response, there are a group of thirty proteins called complements and another group called interferons that circulate in blood. Complements destroy bacteria by attaching to them then signaling for an inflammation to begin. Complements sometimes simply destroy the bacteria by themselves. Interferons focus on viruses and prevent them from replicating once inside the body.

It seems a pathogen doesn't have a chance. But many fight back as shown by the coagulase-producing staphylococci. More ingenious pathogens such as the AIDS and herpes viruses and tuberculosis bacteria avoid immune reactions by hiding inside the body's own cells.

The AIDS virus may be the most insidious foreigner that infects humans because it enters cells of the immune system itself. HIV attaches to and enters T-cells of the lymph system. It then mixes its genes with T-cell DNA. Instructions for producing thousands of new viruses are made inside the T-cell. By hiding inside human cells, HIV avoids destruction by the host's defenses against foreign particles.

Recognition of "foreign" is the first step of the immune response. HIV and other viruses such as influenza mutate frequently. The mutations often result in subtle shifts in the composition of the virus' outer surface. Recognition of a foreigner then becomes a dicey task for leukocytes, and the invader gets the upper hand.

In tuberculosis the immune system works as it should, but with a deadly complication. Special immune cells called macrophages "eat" *Mycobacterium tuberculosis* by enveloping it. But while the macrophages prepare to digest the *M. tuberculosis,* they concurrently carry the pathogen in the bloodstream to the lungs. Lung tissue is the pathogen's target tissue. The bacteria that escape decomposition in macrophages begin multiplying, and the disease symptoms begin.

HIV, herpes, and TB are just a few examples of human pathogens that have evolved methods to avoid or to use the immune system for their own purposes. It's not surprising that these diseases remain particularly difficult to eradicate.

Acquired Immunity
Emil von Behring was awarded a Nobel Prize in 1901 by showing that immunity can be transferred from one animal to another. This is accomplished by the body's capacity to produce **antibodies. Acquired immunity** is the process of developing an antibody specific to a certain invader in the body, particularly when the invader is reinfecting. Mothers provide some antibodies to the fetus before birth, and there is also an artificial means of acquiring immunity. It is vaccination.

Like innate immunity, acquired immunity reacts to foreign particles shortly after they enter the body. Part of the leukocyte response includes signaling the immune system to make antibodies when they are needed in the bloodstream or on the skin. The very first time a pathogen infects, antibodies help out to defend by joining the team of leukocytes, lymphocytes, complements, and interferons. More valuable than that, however, is the role antibodies play in reacting to reinfection. In a number of instances, antibodies can react to reinfection even if the pathogen returns years later.

Antibodies

Antibodies are proteins that bind with specific molecules on a microbe's outer surface. Once bound, the antibody holds on tightly but doesn't harm the microbe. Rather it sends a signal for specific cells to arrive and complete the destruction of the microbe.

After an antibody initially contacts a microbe, the immune system learns how to produce more of the same antibody, quickly. The next time the microbe appears, the body is prepared to attack it. Antibodies are designed so that they bind to features that are characteristic of the infectious agent, called an **antigen**. It is the antigen that prompts the immune response that next time around.

Think of the innate immune system as a SWAT team dispatched at once to seek and destroy an unknown enemy. Sensing that the trespasser may return again and again, additional forces (acquired immunity) are trained in special techniques (antibodies) to conquer it. Acquired immunity takes longer to spring into action than innate immunity, which keeps a constant watch over its domain. Innate immunity reacts within minutes. Acquired immunity's antibody response takes longer but it gets better with practice. When a foreigner is recognized because of previous infection, the body can produce high levels of the needed antibody quickly, within days. Over your lifetime, acquired immunity will be your main protection from a number of recurring diseases.

Vaccines and the Vaccination Question

Vaccines provide acquired immunity. A vaccine is a suspension of antigens, and when injected into the bloodstream, it induces the body to start making antibodies. Successful vaccination is achieved when the mere presence of the microbe is detected, recognized, and the immune system immediately gets to work producing the corresponding antibody. Vaccination is a preventive measure. The world offers no absolute certainty that an adult or child will someday be exposed to any given disease.

The major viral diseases that we vaccinate against in the United States are measles, mumps, chickenpox, rubella, hepatitis B, and polio. Hepatitis A and rabies vaccines are also available for people who may have a high risk of exposure. Yearly vaccination programs also target the new strains of influenza viruses, sometimes with mixed results.

In general, diseases caused by viruses are *prevented* through vaccination. Bacterial diseases tend to be *treated* with antibiotics after infection begins. The few vaccines for bacterial diseases include vaccines against tetanus, diphtheria, and whooping cough. The reason vaccines are developed for viruses more often than for bacteria is due to the composition of the viral outer surface compared with bacterial outer walls and membranes. Virus exteriors are high in proteins. It's easier for the immune system to form antibodies to proteins than to the complex carbohydrates that predominate on bacteria.

There are different types of vaccines depending on how they are made. Our current arsenal includes (1) attenuated whole vaccines made from live pathogens that have mutated over time, resulting in the loss of virulence, (2) inactivated whole vaccines made from killed pathogens, (3) subunit vaccines made from pieces of the pathogen that will stimulate an antibody response, and (4) toxoid vaccines made from toxins that have been detoxified by heat or chemicals.

Biotechnology companies are developing new recombinant

vaccines made from a blend of microbial components. Also in the works are DNA vaccines. In these, injected DNA carries antibody instructions for a specific pathogen. The host DNA must take up the new DNA, and then reproduce it. If it works as planned, the host will forever make the desired antibodies.

Is Vaccination Safe?
Physicians follow a recommended schedule for childhood immunizations. In Western medicine, the program begins at two months of age and goes through eighteen months, followed by boosters in later years. Flu shots are given yearly.

There is no small amount of disagreement on the benefits and the hazards of vaccination. Smallpox has been eradicated, so vaccine proponents use this example to advocate for further immunizations against other diseases. The eradication of smallpox was truly a shining moment in medical history. The success against smallpox may have led quite a few people to think polio, whooping cough, and even measles have also been defeated. Physicians have mistakenly assumed polio is nearly eradicated because they have never seen a polio patient. In fact, polio cases worldwide have been on an increasing trend for the last six years. Measles is decreasing worldwide, but over 500,000 children under age five still die from it each year.

The case for vaccination uses part logic and part statistics. If enough people, principally children, receive vaccine for a given disease, further vaccination to cover the entire population is unnecessary. This is due to **herd immunity**. Health-risk specialists have calculated the percentage of people that must be infected by a pathogen before infection will spread. Various diseases have their own percent value. There is also a portion of any population who are already immune to the new infection. The percentage of immune people eventually reaches a type of critical mass in which the probability of transmission becomes very small. The community, or the herd, acts as a single living thing, not unlike an ant colony. If a colony is 80 percent immune, for example, then the remaining 20 percent are fairly safe from contracting the disease. There is safety in the herd. A vaccination program must achieve the 80 percent level of immunity within the herd.

Herd immunity can be a tricky concept to accept. It is based on statistics rather than individual choice. Furthermore, it's influenced by the blend of susceptibility and opportunity within the population plus the virulence of the pathogen. These are not things you can see as you walk down the street and view your fellow citizens. Herd immunity requires having faith in the statistics, which is understandably difficult for a parent with a young child whose doctor says it's time to start vaccinating.

Increasing numbers of people question or oppose vaccination, and they use the herd concept as ammunition. They argue, "If doctors haven't seen polio or measles, then why should parents worry about these diseases? Herd immunity will protect the rest of us who aren't vaccinated." The fatal flaw in the argument is today's global community. Herds lose cohesiveness when new members constantly enter the herd and others leave. Humans no longer mimic ants' independent and isolated colonies. In humans, individual behavior reigns. Immigrants or travelers who enter new regions may be immune to a different set of diseases than are common in the United States. Germs travel great distances these days. The transit of a virus in Marco Polo's time took years. Today the SARS virus can move halfway around the globe in hours. Herd immunity still exists, but the points where it may be disrupted have become numerous.

Opponents of immunization also argue that vaccines are simply not safe. Vaccines from attenuated live viruses are those made from viruses that have been rendered harmless through mutation. Vaccination opponents pose the question, "If a mutation can get rid of virulence, then another mutation can cause virulence to return." Though this occurrence is rare, it is thought to be possible. It is estimated that the polio vaccine presents a one-in-a-million chance of causing polio.

Vaccines can be made from partially inactivated toxins, also causing concern. Just the idea of a toxin that is "partially" inactivated will lead many people to decide the risks from the vaccine are too great.

Vaccinations with live virus are used for some diseases, that is, measles. It is known that vaccination with live virus often causes side effects. Up to 15 percent of recipients of measles vaccine develop a rash, fever, or joint pain.

When live viruses are destroyed before being put into a vaccine, the chemicals used to destroy them include phenol and formalin. Both chemicals are toxic to humans. In addition, viruses must be maintained in living cells in order to multiply to levels needed for a vaccine. Flu vaccines are made by propagating the virus in chicken egg embryos. Egg proteins then become part of the vaccine suspension and may potentially cause allergic responses in recipients who are sensitive to eggs. There is scientific debate on this subject. Proponents of vaccination suggest the amount of egg protein in the vaccine is not enough to cause a severe allergic reaction, yet CDC publications strongly recommend against giving certain egg-based vaccines to people prone to anaphylactic reaction to egg protein.

Vaccine opponents don't stop with those arguments. They present evidence linking vaccines to serious disorders in humans. Medical and university studies have accumulated data proving and disproving the idea that vaccines cause acute syndromes. The vaccine-disease links most often debated are: (a) MMR (mumps, measles, rubella) vaccine and autism in children, (b) hepatitis B vaccine and multiple sclerosis, (c) polio vaccine and HIV transmission, (d) rotavirus vaccine and intestinal blockage, (e) varicella (chickenpox) vaccine and the spread of chickenpox to other family members, and (f) pertussis vaccine and seizures.

Each side of the vaccine controversy uses extensive analyses to accumulate data weighing in their favor. Many people realize that statistics can be used to show what a person wants to show. Almost all studies published by the medical establishment go through critical review by scientists trained in their field and the statistics cited in medical journals are scrutinized. As for many of the controversies in science, it is prudent to conduct a thorough investigation of the subject from reputable sources.

Colds and Flu

It doesn't seem there could be much more to learn about colds and flu. Yet they are often confused with each other. People also mistakenly believe colds and flu are caused by bacteria rather than viruses. Misconceptions have been around a long time. The term influenza dates to the Middle Ages, when certain positions of the stars were thought to "influence" the onset of the disease.

The Common Cold

Rhinovirus and human coronavirus are the principal cold viruses.

Rhinovirus is one of the smallest of the viruses—about 25 nanometers in diameter (one forty-millionth of an inch). Its preferred temperature is 91.4–95 degrees F (33–35 C), about the same as the temperature inside nasal passages. There are at least one hundred different **serotypes** (a specific trait that induces antibody production) of rhinovirus. Limitless combinations of serotypes allow colds to occur repeatedly in the absence of a satisfactory immune response. The body simply cannot make sufficient antibodies to defend against all the permutations.

Coronavirus is famous as the cause of SARS, but is overlooked as a cold virus. The round virus is 80 to 160 nanometers in diameter, and has clublike projections covering the outer surface and making it resemble a crown, or corona (Figure 1.7). Like rhinovirus, its main mode of transmission is by way of fomites. Sneezing and coughing are secondary modes.

Commoncold, Inc. has tackled a number of myths regarding colds. Some of these may soon be de-mythed! Recent studies on "feed a cold, starve a fever" are beginning to suggest the old maxim may have a bit of merit. Without conclusive evidence, these are some of the prevalent common-cold myths:

Feed a cold, starve a fever. The medical community overwhelmingly agrees that it's best to ignore this maxim and get plenty of nutrients, fluids, and rest whenever you have a cold or a fever.

Susceptibility to colds needs a weakened immune system. In both the healthy and those weakened with illness, once a cold virus enters the nose, it almost always infects.

Heat that dries mucous membranes help in catching a cold. Though the nose feels dry, the protective mucous membranes maintain their integrity in low humidity.

Getting a chill leads to a cold. Studies on chilled versus nonchilled volunteers who were inoculated showed there was no difference.

Cold symptoms help you get over the cold faster. Once the virus infects the nasal passages, symptoms begin, but the cold runs its course as viruses multiply inside cells. Sneezing and runny excretions help the virus by spreading it other people.

Drinking milk increases nasal mucus during a cold. Milk is digested like other foods and doesn't cause mucus buildup.

Influenza

The influenza virus (flu) was first identified in the 1930s. Influenza viruses are now grouped into A, B, and C types based on the molecules that project from the virus coat and act as antigens. Influenza Type A and its subgroups are commonly associated with outbreaks in human populations. Type B causes lesser outbreaks, and its symptoms are milder than Type A's. Type C is at this time not considered a threat to human health.

New vaccines are prepared against the flu each year because the virus mutates frequently, and so last year's version of the flu is dissimilar to this year's. The body's immune system cannot recognize the different sets of viral antigens presented to it each year and therefore does not produce sufficient antibodies to attack the virus. The complete assortment of mutations that make each year's flu virus look different to our immune system is called an **antigenic shift**. Antigenic shift in influenza may be viewed as an evolutionary adaptation that allows it to persist in a population, claiming hosts repeatedly over decades. Compare influenza's ever-changing composition to a virus such as measles. A measles vaccine and booster confers immunity to almost 100 percent of the people who receive it.

Different flu versions may come from a different animal reservoir each year. The main reservoirs are birds, pigs, and horses. Flu outbreaks often trace back to rural areas in China where human and animal populations may live in close association. Birds are an important reservoir for the influenza virus, and bird flus have contributed to the notable outbreaks in recent decades. To make vaccine for each new season, the new version of the virus must be identified early in the year. Hundreds of strains from sources around the world are collected and analyzed, and then the most probable mix of strains is selected. A few strains may be selected, or, in some years, a dozen or more are combined to develop the vaccine.

Development, production, and distribution of vaccine drugs take several months. If the strains aren't selected with care, vaccine for the upcoming flu season will be ineffective. Scientists prefer not to admit that they make guesses. They use all their knowledge of virulence, transmission, epidemiology, and health trends to decide which strains may cause an impending threat. After all the hard science, there remains a small piece of the pie left to speculation (guessing). Even so, they usually get it right. Vaccines are generally effective in preventing each year's flu virus.

Unlike rhinovirus, influenza doesn't stay in the upper respiratory tract. It works its way along the cells lining the deeper respiratory tract to the lungs. Body aches, chills, and fever are the common symptoms. Many people also experience coughing, sore throat, and headache. Aches, chills, and fever are the result of your immune system doing its job in arresting and destroying the virus. Flu does not cause stomach upset. The term "stomach flu" describes a real illness, but it is caused by the Norwalk viruses and not influenza.

Direct harm from the flu virus comes from toxic products released as the virus migrates through the respiratory tract. When the toxins pass into the blood, they cause symptoms in addition to the symptoms from your immune response to the virus: dizziness, headache, fatigue, and fever and chills.

The very young and the very old are at the highest mortality risk from flu. The lungs inflame and other microbes take the opportunity to start their own secondary infections. All through these events the nascent immune system of the newborn or the weak immune system of an elderly person cannot cope with the microbial onslaught. Influenza and pneumonia together amount to the most deaths in the United States from infectious disease, particularly among the elderly.

Spreading Colds and Flu
Colds and the flu are connected to cool, rainy conditions that in North America occur from late fall to early spring. There is no evidence that rhinovirus is more prevalent in the environment during this period, though it is thought that cooler temperatures and moist conditions help preserve it. Flu follows a more cyclical schedule than the common cold. The emergence and recession of flu may be due to the breeding cycles of its animal reservoir rather than the effects

of climate. If there is a link between cold wet weather and increased infections, the reason may be explained by simple human behavior. People tend to crowd indoors when it's cold and wet.

Hand-to-face touching transmits rhinovirus; the infectious dose may be as low as a dozen particles. Cold symptoms come within two to four days and include mucus secretions, watery eyes, and sneezing. By contrast, Influenza A is better transmitted in aerosols, and it usually needs a higher dose (thousands to millions) to make you sick. Flu symptoms in otherwise healthy people start in from one to three days, but virus particles can be shed from an infected person for a day or two before symptoms are apparent.

Staying home from work to nurse a cold or the flu is the best way to prevent transmitting it to coworkers. Unfortunately, staying home is a considerable challenge for some. The National Foundation for Infectious Diseases (NFID) found that, in 2005, 35 percent of U.S. workers felt pressured to go to work when they were miserable with the flu, even though almost half said they were annoyed by sick coworkers. What could be the reasons for dragging yourself into work when you're sick? The NFID dissected the psychological pressures of tending to an infection at home. Sixty percent of workers were concerned about getting their work done; 48 percent felt guilty about staying home; 25 percent were not paid for sick leave; 24 percent said their employer did not allow sick leave, or gave very few days for it; and about 20 percent each said that the boss would be angry or they could lose their job. In a typical year, employees who come into work when they have the flu collectively amount to $10 billion in lost productivity.

Echinacea

"Home-brewed" or natural treatments for colds and the flu are gaining more and more shelf space in drugstores. Popular ingredients are echinacea, vitamin C, zinc, manganese, potassium, various

B vitamins, and amino acids. The FDA will approve a **drug** only after it has been tested for safety and effectiveness in animals, then small groups of people, and then thousands of people. "Natural" cold remedies not do receive such testing, so the FDA will not allow words like "treatment," "cure," "drug," or "disease" to appear on the label. They are labeled dietary supplements, which are subject to more lenient FDA requirements.

Most health agencies in the United States report there is no evidence supporting echinacea as a remedy for cold and flu symptoms. Conversely, some studies suggest it boosts the immune system and reduces inflammation. Echinacea plant extracts contain compounds thought to have therapeutic effects: protein-carbohydrate complexes, oils, and antioxidants. These types of compounds are known to work in combination with the immune system to affect wound healing and inflammation. Despite the logical reasons for including echinacea in anticold products, there is scant definitive evidence that it acts directly on viruses or bacteria.

Western medicine is attached to the idea that if a compound hasn't been approved by the FDA, then it doesn't work. The world is a big place. Nonwestern cultures depend on diverse medical treatments without the slightest thought of FDA testing and have done so for generations. Perhaps a blend of technology-driven Western medicine with traditional nonwestern forms of prevention, treatments, or cures will prove to be the most successful approach.

Bird Flu

The H5N1 influenza strain is avian influenza or bird flu. Its reservoir is thought to be wild birds, which in turn infect poultry. Like other flu viruses, a mutation allows it to "jump" from animals to human hosts, though the jump of H5N1 from birds to humans is rare. The world health community's current concern is the probability of an H5N1 flu **pandemic**. (An epidemic is a spread of disease to large

The Great War

In March 1918 war-weary American soldiers returned home from mud-choked trenches and frozen forests. The Great War, which came to be known as World War I, neared its end. Troop movements and call-ups continued as combatants on both sides of the Atlantic prepared themselves for the conflict's final months. Survivors of the battles across Europe breathed a sigh of relief as they boarded ships and pictured their families waiting for them on farms and city streets across the continent. As thousands of troops returned home, wary replacements mustered out, praying for the Armistice.

On a chilly spring morning at Fort Riley, Kansas, a young private stepped out of breakfast formation and walked into the camp's military hospital. His complaints of fever, sore throat, and headache were echoed by a hundred more soldiers by lunchtime. Within a week, five hundred were admitted with symptoms, and by the end of the spring, forty-eight had died. Doctors, families, and politicians at the time did not realize the horror that would befall America in that year. By the end of 1918, nearly 700,000 people died in the United States from the Spanish influenza pandemic that also ravaged Europe, Asia, Africa, Brazil, and the South Pacific. By 1919, the flu would claim 50 million lives worldwide.

From March onward, sick soldiers and sailors still in uniform entered stateside hospitals. Their families soon caught the flu, first those living in cities, and then, inexorably, the epidemic spread into the countryside and westward. By September, a man who had recently moved from Chicago to San Francisco fell ill. In that one month, 12,000 Americans died. October's total reached 195,000. As death enveloped the country, fear of the evils of bioterrorism soon emerged. One anxious health official disclosed a theory that was prophetic of our present commingling of science and nationalism. He surmised, "It would be quite easy for one of these German agents to turn loose Spanish influenza germs in a theater or some other place where large numbers of persons are assembled. The Germans have started epidemics in Europe, and there is no reason why they should be particularly gentle with America."

Through the summer and fall, hospitals were full, and the dead were stacked in corridors leading to the morgues. Trucks rattled through city streets, stopping repeatedly to pick up caskets and corpses. Microbiologists took stabs at developing vaccines, but hampered by their limited technology of the times, they incorrectly presumed the disease was caused by bacteria rather than a virus. Each new vaccine failed. Surgeon General of the Army Victor Vaughan helplessly concluded, "If the epidemic continues its mathematical rate of acceleration, civilization could easily disappear from the face of the earth within a few weeks."

While doctors frantically sought treatments or preventions, cities and towns took steps to protect the few citizens that hadn't yet been caught in the epidemic. Schools and theaters closed. Police wearing face masks—"Flu Squads"—dispersed crowds and shooed pedestrians off the streets and arrested those who weren't wearing masks. When Armistice Day came in November, an upwelling of parties and parades served to begin a new spread of the disease.

The streets were barren by the time Christmas came. The few survivors of the disease and the lucky ones who avoided infection had learned their lesson about germs spreading through the air and in crowds. The few stores open for business skipped their holiday sales to discourage crowds of shoppers. Office and factory workers stayed home. Children, always gifted at finding the slightest ray of optimism, jumped rope to a new rhyme that had spread across the country hand in hand with the scourge:

I had a little bird,
His name was Enza.
I opened the window,
And in-flu-enza.

At the start of the new year, the Spanish flu pandemic ended. No action from the medical community seemed to have played any role in the disease's disappearance. A public health official in Los Angeles opined that the infectious agent had simply been too virulent for its own good. "It ran out of people who were susceptible and could be infected," she said.

The 1918-1919 pandemic killed more people than had been killed in the entire world war. It was eventually learned that Spain had never been the origin. Scientists traced the probable origin to China, where a genetic mutation created a unique and highly virulent strain so new to the population that no herd immunity existed against it. Virtually every person exposed to the 1918 flu virus was susceptible to it.

In the 1990s, a team of pathologists recovered 1918's flu virus DNA from a victim immersed in the Alaskan permafrost. They characterized an unusually proficient system within the virus for binding to mucosal cells, thus explaining its extraordinary virulence, yet failing to explain the impetus for the pandemic. Under the cloud of present-day bird flus, it may be unsettling that many virologists today offer the same words spoken by scientists who by the end of 1919 had put hands on hips and wondered, "What happened?"

numbers of people in a given location, such as a country, island, or continent. A pandemic is a larger epidemic that spreads around the globe.) The likelihood of a pandemic increases if H5N1, or any bird flu, and an existing human flu virus simultaneously infect a person and the viruses swap genes. The resulting new virus could carry the virulence of bird flu *and* be transmissible among people. H5N1 has yet to fulfill the dire predictions made for it in the past few years. Increasing global travel may, however, aid the fast spread of virulent flu strains worldwide in the near future, perhaps leading to a pandemic.

Some antiviral drugs are available in limited supply as treatments for a small number of flu viruses. Their drawback is that they work only after infection begins. Treatment drugs interfere with gene transfer between the virus and host cells. Oseltamivir (Tamiflu) and zanamivir (Relenza) are the current antiviral treatments for bird flu. Vaccines remain more valuable than treatment drugs because vaccines are *preventive,* so they function in halting the spread of infection before it affects an entire community.

Sexually Transmitted Diseases (STDs)
STDs are caused by bacteria and viruses, yeasts, and protozoa. Transmission is efficient; the host conditions are welcoming; and there exist behavioral traits in society that make people avoid seeking treatment. STDs are a pathogen's dream come true.

Several factors complicate the estimates of STD incidence rates. Many incidence tables published by health agencies sort their data based on age groups, current health, or HIV status. STDs are complicated by the prevalence of multiple infections leading to more than one disease at a time. Some STDs, such as herpes and AIDS, persist for years before causing symptoms, making the number of new cases difficult to calculate. Methods used in reporting incidence to state health agencies are inadequate. Though eight billion dollars are spent annually to diagnose and treat non-AIDS STDs, only four must be reported by diagnosing doctors to health agencies: gonorrhea, syphilis, *Chlamydia,* and hepatitis B.

Notwithstanding the difficulties in tracking STDs, the American Social Health Association has compiled the following statistics:

- The number of people in the United States with viral STDs is estimated at 65 million.
- There are 15 million new STD cases each year.
- One in four adults has genital herpes; the majority do not know they have it.
- One in four teens contract an STD each year.
- Over 50 percent of all new STD cases are in people between the ages of fifteen and twenty-four.

In the United States, gonorrhea (caused by bacteria), syphilis (bacteria), and genital herpes (virus) are prevalent. Worldwide, AIDS is a leading cause of mortality. Other STDs are nongonnococcal urethritis (NGU, caused by *Chlamydia* bacteria), Candidiasis (*Candida* yeast), genital warts (human papillomavirus), and trichmoniasis (*Trichomonas* protozoa). Hepatitis B is an often-overlooked STD though it's one hundred times more infectious than HIV. Like HIV, it is transmitted sexually, through direct blood transfer such as needlesticks, and from mother to fetus. Up to 30 percent of infected people may be unaware that they have hepatitis B.

The normal flora of female genitalia change with age. Lactobacilli bacteria predominate, and as they grow they produce lactic acid, and the acidity prevents other microbes from growing in the area. When conditions change to cause a decrease in lactobacillus, other microbes that were kept in check take advantage of the opportunity to grow. Pregnancy and menopause are associated with a drop in lactobacillus numbers, and the incidence of yeast or protozoa infections tend to increase in these conditions.

STD microbes are not found on household surfaces. Because STDs require direct contact between individuals for transmission, they tend not to persist outside the body. STDs have one important point in common with other infectious diseases. The rate of antibiotic resistance is accelerating.

Antibiotic Resistance

The March of the Resistant Microbes

Antibiotic resistant microbes began their stealthy inroads into the medical community not long after penicillin was commercialized in the 1940s. *Staphylococcus aureus* was the first organism thought to have developed permanent resistance to penicillin. This seems logical in retrospect. *Staphylococcus* is ubiquitous in the environment. No matter the disease for which penicillin was prescribed, there was plenty of *Staphylococcus* on each patient and on their bedsheets, pajamas, and personal items. Generations of bacteria in close association with antibiotic-treated patients had the opportunity to develop and refine their mechanisms for destroying penicillin and other antibiotics. In less than thirty years from the first use of penicillin, resistant *S. aureus* emerged.

Resistant microbes were noted as a curiosity by many in health care for decades. The list of "problem" microbes was short: *Streptococcus* pneumonia, *Enterococcus* intestinal infections, and resistant gonorrhea strains. Into the 1980s, doctors continued to prescribe antibiotics as a safety net for seemingly any and all ailments. If occasional resistant species protracted the treatment course, a number of alternate antibiotics were available from which to choose. If resistance happened to develop against those alternatives, many reasoned, medical technology would discover new sources in far-off jungles, or talented chemists could synthesize an indestructible new drug. Toward the end of the eighties, certain perceptive people began to understand that the rate of bacterial adaptation to antibiotics was outstripping the speed with which new ones were developed. By 1994 the *New England Journal of Medicine* reported a bacteria sample from a group of patients in which the microbes recovered were resistant to *every* antibiotic that was currently available in Western medicine.

Selecting for Resistance

An antibiotic is a compound made by bacteria or fungi that prevents the growth of other microbes. The main producers of antibiotics are the bacteria *Bacillus* and *Streptomyces* and the fungi *Penicillium* and *Cephalosporium*. By producing antibiotics to ward off other species, these microbes can hold their territory and have unhindered access to precious nutrients. Antibiotics are made inside the cell, then excreted into the surroundings; they are **extracellular** compounds. If the antibiotic were to remain inside, its defensive benefits would disappear.

There are a few synthetic antibiotics currently in use. Sulfonamides or sulfa drugs are an example. One trick in building an effective synthetic drug is to ensure a potency that kills pathogens but doesn't cause massive harm to the patient.

Antibiotics kill bacteria in a variety of ways, principally by stopping reproduction. For instance, penicillin disrupts the steps bacteria use to build a new cell wall. Other antibiotics interfere with either chromosomes or protein production systems. Still others damage the bacterial membrane that lies inside the sturdy cell wall. No matter the mode of action, an antibiotic is configured in nature to kill unrelated microbes and leave unharmed the producing species.

Bacteria develop resistance to antibiotics through generation upon generation of spontaneous mutations. A random mutation occurs on a gene in one bacterial cell. The mutation gives that single cell an attribute, allowing it to withstand the effects of an antibiotic. It happens by chance. The cell multiplies to millions quickly due to its fast growth rate. At the same time, bacteria exchange DNA. The new resistant cells exchange small snips of DNA with nonresistant cells. If the piece of exchanged DNA contains the vital gene for resistance, the percentage of resistant cells begins to increase.

An antibiotic in your bloodstream will successfully kill all the "normal" nonresistant bacteria. But the antibiotic also "selects" for bacteria that survive its killing action. The antibiotic may make target bacteria *more* resistant. These resistant populations are known to inhabit hospitals, nursing homes, and sometimes outpatient clinics and child care centers.

Plasmids, small strings of DNA separate from the bacteria's main piece of DNA (its chromosome), have been found to carry more than one gene for resistance. Therefore, one strain can be resistant to more than one antibiotic. Bacteria pass plasmids back and forth, making the spread of resistance in a species and between species even more prevalent.

Resistant bacteria can destroy an antibiotic before it destroys them. Some bacteria produce enzymes that cleave antibiotics. Others change the makeup of their outer surface just enough to prevent the antibiotic from binding to it. Some "superbugs" contain pumps, called antibiotic efflux pumps, in their membranes. These pumps purge the cell of the antibiotic as soon as it gets inside. These very same pumps are thought by many to eject chemical disinfectants from the cell, making the bacteria resistant to disinfectants as well as antibiotics.

The Antibiotic Business

That's the How. We are more concerned with the Whys of antibiotic resistance. For years antibiotics were overprescribed, prescribed for the wrong infectious agent, and misused. An unfortunate result of penicillin's emergence as a "miracle drug" is the notion that a pill will solve all health problems. Penicillin and then other antibiotics were given to patients who probably would have gotten better on their own with bed rest. Perhaps the antibiotic was prescribed for a viral infection. In addition, once a bottle of pills is taken home, people often stop taking them as soon as they feel better. All these factors contribute to the proliferation of resistance, even in bacteria with marginal capacity to become resistant.

Antibiotic use in the United States is massive. Over 50 million pounds of drug are produced each year. Agriculture accounts for 40 percent of the use; antibiotics are given in feeds to cattle, pigs, and poultry for improved growth rates. Additionally, antibiotics have been used in spraying fruit trees to prevent infections. Articles on the extent of resistance transferred from food to humans might reach to the moon if stacked in a pile.

Many countries outside the United States sell antibiotics without a prescription. This practice leads to overuse, underuse, treatment for inappropriate ailments, consumption of outdated antibiotics, and use of an uncontrolled substance that may be adulterated.

Antibiotics are put into different classes based on their chemical structures, but it's more important to know the viruses and bacteria that are destroyed by a given antibiotic. The chemical names in Table 6.2 go by trade names at the pharmacy. Read the information sheet inside the drug's packaging or the instructions supplied by your pharmacist.

The major resistant bugs are methicillin-resistant *Staphylococcus* (MRSA); vancomycin-resistant *Enterococcus* (VRE); multiple-antibiotic-resistant *E. coli;* penicillin-resistant *Streptococcus pneumoniae;* and multidrug-resistant tuberculosis. An emerging resistant TB strain is called "extensively drug-resistant TB," or XDR-TB. This microbe is a virulent and extremely dangerous health threat in South Africa among HIV-positive patients. Close to 100 percent of HIV-positive and AIDS patients in that country have died from TB caused by XDR-TB. The strain is resistant to all known first-line defenses against tuberculosis and also at least two second-generation antibiotic classes. At this time, XDR-TB has been detected in thirty countries.

Common Antibiotics

Antibiotic	Kills
Penicillin G or V	Gram-positive bacteria
Oxacillin	Bacteria resistant to penicillin
Ampicillin	Broad spectrum that kills Gram-positive and -negative
Amoxicillin	Broad spectrum
Cephalothin	Gram-positive bacteria
Bacitracin	Topical application for Gram-positive bacteria
Vancomycin	Gram-positive bacteria
Chlorampenicol	Broad spectrum
Streptomycin	Broad spectrum, includes *Mycobacterium* (TB organism)
Neomycin	Broad spectrum topical application
Gentamicin	Broad spectrum
Tetracycline(s)	Broad spectrum, including *Chlamydia*
Rifampin	Mycobacterium
Ciprofloxacin	Broad spectrum for urinary tract infections

Table 6.2. Today's most-prescribed antibiotics are broad spectrum.

Some drug treatments that are not antibiotics may also be thwarted by resistant microbes. Two examples are the drugs Zidovudine (AZT) for the AIDS virus and Acyclovir for herpes virus.

Fungal Diseases
Types of Mycoses
A fungal infection is called a **mycosis**. It seems fungal infection and disease are treated as poor cousins to those caused by bacteria and

viruses. On the contrary, mycoses are chronic problems affecting millions of people. Mycoses can be difficult to diagnose and to treat. Many fungal skin infections have similar symptoms, and considering there are over 100,000 different known fungi in the world, it's easy to understand the challenge of accurate diagnosis. General practitioners are often not experienced in distinguishing the fine points of similar rashes caused by unrelated species. Dermatologists do have the required expertise, but because they are medical specialists, many underinsured people might not seek their opinion. Furthermore, fungi grow slowly in skin infections, so a mycosis can progress for some time before being treated.

Mycoses are classified according to level of fungal penetration from the outer epidermis into deeper layers, and then organs. Superficial mycoses are on the skin surface or hair only. Dandruff is superficial. Cutaneous mycoses (dermatomycoses) penetrate the outer skin layers or into hair shafts. Examples are athlete's foot and ringworm. Subcutaneous mycoses are in the skin and below and affect connective tissue and bone, such as the lesions of sporotrichosis. Systemic mycoses are infections of organs through spread of the fungus in the bloodstream and are often fatal. Examples of fungi causing systemic infections are: *Histoplasma, Fusarium, Blastomyces,* and *Coccidioides.*

You may be most familiar with the cutaneous mycoses, tinea pedis and tinea cruris from using their common names, athlete's foot and jock itch, respectively. Products sold over the counter for these ailments treat the symptoms only and do not eliminate the fungus. The general symptoms of cutaneous mycoses include itching, redness, rash, lesions, pustules, and more serious deformations. The main antifungal drugs used for cutaneous mycoses are Amphotericin B, ketoconazole, miconazole, naftifine, and griseofulvin. Tolnaftate is frequently used for treating athlete's foot.

Candida albicans yeast infections can be cutaneous or systemic. The yeast is a unicellular structure unlike the filamentous fungi

(fungi that grow by building thin filaments that elongate and therefore spread over a surface or into a porous, material-like skin). *Candida* causes the oral infection thrush and vaginal candidiasis. Cutaneous *Candida* infections are treated topically with miconazole or clotrimazole. In AIDS patients, opportunistic *Candida* may cause an aggressive and possibly fatal systemic infection.

Mycotoxins

Fungi don't always cause a defined set of symptoms because their modes of action are diverse. Dermatophytes, fungi that infect the skin, act as parasites by eating through the skin as they grow. Other fungi also produce toxins. Any toxin produced by a fungus is called a mycotoxin. When ingested, mycotoxins cause severe digestive upset, and if the toxin reaches the bloodstream, neurological damage occurs.

The plant pathogen *Claviceps purpurea* produces ergot, a strong hallucinogen. When *Claviceps* grows on plants such as rye, spring wheat, or barley, the harvested grain may contaminate food, with hazardous results. The Salem witch trials in the 1600s are thought to be the sad outcome of ergot poisoning in many of the citizens. Victims exhibiting disorientation and erratic behavior had likely been suffering the toxic effects of eating contaminated grain. A generation or two later, ergot became a source for making LSD.

Aflatoxin produced by *Aspergillus* mold occasionally contaminates food: peanuts, corn, and tree nuts such as Brazil, pecan, pistachio, and walnuts. Aflatoxin poisoning is associated with acute necrosis and cirrhosis of the liver and liver cancers.

Do Pathogens Visit or Do They Move In?

Once a pathogenic microbe has invaded your body, will it leave of its own accord if not treated? Most infectious agents leave the body. Notable exceptions are HIV and herpes virus, which may stay forever once they infect.

People often wonder why the deadliest viruses on earth ensure their own demise by killing their host. Why then don't pathogens eventually disappear from our population? It is because infectious agents exit the body. They may leave behind a corpse, but the survival of their species is ensured by using preferred **portals of exit.**

The portals of exit that pathogens use are related to their entry routes. A cold infects the mucous membranes, and by riding along in sneezes and runny discharge, it finds a way to move on to the next person. Exit routes other than mucous secretions include pus and discharges from wounds, urogenital secretions, and . . . you can figure out the rest.

Summary

Infection and its transmission through communities is a science built on shifting sands. Pathogens evolve over time. Their resistance to drugs evolves, too, and they can respond to almost any medical invention before doctors realize that a new threat is here. In addition to our innate immunity, there are techniques available for stopping the spread of germs before an infection takes root. Some of these techniques are as simple as washing hands and being mindful of the things we touch. Chemicals and drugs may be more sophisticated than soap and water, but it might not matter. Pathogens have shown that they are more than capable of meeting any weapon we throw at them. The trick is to stay one step ahead, even if it is a very small step.

7

These Are Microbial Times

Hey the line forms, on the right dear
Now that Macheath's back in town.

—Bertholt Brecht, Marc Blitzstein

Human behavior is the greatest determinant of whether infection occurs in a population and how quickly it spreads. Our decisions influence the progress of an infectious agent once it has made entry into our community, and our actions will help or hinder its transmission thereafter. We control the outcome of infectious disease more than the pathogen's virulence itself.

There are a few success stories in which a threatening disease has been eliminated, but there remain hundreds of infectious agents that continue to haunt humans and animals worldwide. Perhaps most discouraging are those agents that were thought to be conquered and are now returning.

A well-known virus, a unique bacterial species, and a virulent cousin of a familiar bacterial strain are three examples of the challenges of eradication and the menace of reemergence.

Humans and Their Infections

The history of viruses on earth is part of the history of humans. A virus needs a host cell. Different viruses infect plants, bacteria, and animal cells. The viruses that infect humans sometimes also infect another animal or insect. For many years a chicken-or-the-egg argument flourished in explaining the origin of mammalian cells and viruses. Did cells develop their ability to replicate after a virus infected them? Perhaps a virus remained inside a primitive cell and evolved with the cell as part of its infrastructure. Or did the earliest cells shed pieces of nucleic acid and proteins, giving rise to the virus particle we know today? In 1951, the geneticist Barbara McClintock conducted experiments with the idea that bits of gene-containing DNA could translocate, or jump, to different parts of a chromosome. Her theory was ridiculed, then ignored for decades until the budding science of gene technology, now called biotechnology, showed that genes can indeed move around on the chromosome. This revelation provided the final piece of evidence needed to confirm that viruses did not form by infecting cells but developed by breaking free of cells, carrying a few essential jumping genes with them. In order to proliferate, they need only infect additional cells and use the cellular apparatus to support their replication. The jumping genes were termed transposons. In 1983, McClintock won the Nobel Prize in Physiology/Medicine for her discovery.

An Eradicated Disease

However they evolved, there's one virus that is believed to have existed earlier than human recorded history. It is smallpox. The earliest evidence of smallpox infection was found in Egyptian mummies entombed from 1570 to 1085 B.C.; their faces bore lesions suggesting "pocks" vesicles. One of the first "celebrity" deaths due to smallpox was that of Ramses V of Egypt in 1157 B.C. around age thirty-five. Smallpox reached Europe in 710 A.D. and migrated to the western hemisphere in 1518 to 1522 aboard ships sailed by the Spanish explorer and conqueror Hernando Cortez. The disease is thought to have been instrumental in the collapse of the Aztec and Inca empires, though this point is debated. When the Spanish conquistadors arrived in Mexico, there were 25 million people native to the area. As the disease spread from the

Spanish to the native population, it took its toll on warriors who would have repelled many of the invaders had they not been so weak that they and their rulers began to die off. A century later, the number of natives was 1.6 million, and historians agree that a major portion of the fatalities was due to the pox. In North America, similar fates befell the Huron, Iroquois, and Mohican populations as immigrants from Europe settled.

Although the basics of immunity were known in ancient Athens, it took hundreds of years and trials to develop successful means for inducing acquired immunity in people who had never been exposed to the pox virus. Acquired immunity experiments began with self-inoculation of the healthy with the pus or scabs from someone who had smallpox. In 1717, English aristocrat Lady Mary Montagu used this technique, called variolation, on her children. Her motivation was likely the result of her brother's death from smallpox and her own facial scarring after surviving an infection. Decades later, physician Edward Jenner used cowpox to refine the procedure to be known as vaccination (from the Spanish word *vaca* for cow).

Jenner's breakthrough was the force behind cooperative efforts in the United States and Europe to eliminate "the most dreadful scourge of the human species" through concerted vaccination programs. From Jenner's first attempts in the late 1700s to the mass-produced vaccines of the twentieth century, the smallpox vaccine was delivered to more and more people until it had become universally available. Governments, health organizations, and local programs took on the awesome task of building consensus on tactics to solve a global health threat. The last case of smallpox transmitted through direct human contact was diagnosed in 1977 in Somalia. By 1980, the World Health Organization (WHO) declared that the virus had been wiped off the face of the earth.

Contrast the collective international effort to defeat smallpox with response to the AIDS epidemic. Since the 1980s, AIDS has presented grave medical challenges for finding vaccines and a cure. More troublesome is the degree to which AIDS has become politicized. Perhaps because of religious and political influence on the response to the worldwide AIDS crisis, there is little international cooperation on options for halting and reducing the spread of HIV. Education is inadequate in many places. For example, there are areas in China and central Asia where the figures of new cases are unknown due to the suppression of disease statistics or the refusal to acknowledge the crisis. Even in the "educated" Western countries, a discouraging amount of misinformation and ignorance persists. If smallpox were to reenter society today and be viewed with the same apprehension as was seen with AIDS, we would have a health disaster beyond comprehension.

A Snapshot of AIDS Today
- 38 percent of people in the United States do not know there is no cure for AIDS.
- Over 400,000 people in the United States live with AIDS.
- In the United States, of the people living with AIDS, 43.1 percent are black.
- In the 1990s, new cases decreased in homosexual males of all races and increased among blacks, Hispanics, and women.
- After a brief drop in new cases, new diagnoses have steadily risen from the year 2000 to the present.
- Heterosexual males and females account for 35 percent of all new cases, a 20 percent increase in the past five years.
- Forty thousand new cases are diagnosed yearly in the United States.
- Fourteen thousand (estimated) new infections occur worldwide each day.
- In the last five years, people under age twenty-five have accounted for half of the new cases.

- As of 2005 there are over 40 million people living with AIDS or HIV worldwide.
- About 63 percent of people in the world living with HIV are in sub-Saharan Africa.
- Africa has 12 million AIDS orphans.
- An estimated 2.6 million people in the world died of AIDS in 2006.
- At present, there is no AIDS vaccine.

AIDS is an example of an emerging disease that over time shifted its global demographics when it jumped from monkeys and chimpanzees to humans. The jump was estimated to have occurred in the 1930s. It worked its way through African and European populations slowly. The earliest known appearances of HIV, based on meager evidence from medical case reports, were as follows: in 1959 in an adult male in the Democratic Republic of Congo; in 1969 in an American teenager who died in St. Louis; and in 1976 in a sailor from Norway.

The urban epidemic exploded in the Western Hemisphere through the 1980s, and then in 1995 the death rate peaked. In this country, the number of new diagnoses is declining due to availability of antiretroviral therapies, but globally, AIDS is pandemic.

A Reemerging Disease

For the first time in history, new tuberculosis infections may be declining, yet it is a reemerging disease. It is defined as reemerging because new cases are thought to be under control, but the disease is far from eliminated and new cases are expected to increase again. Since 1990, the rate of TB infections has decreased slightly in North America, but it has increased in Africa by 130 percent. Like smallpox, TB has been with humans for a long time; pathological signs of consumption (another name for TB) were found in mummies from 2400 B.C. Effective TB chemotherapy was slow in developing, but clinicians in the 1800s seemed to understand the value of breaking the chain of infectious transmission. The idea of sanatoriums as a "cure" brought a stigma, but sending TB patients to them achieved results by effectively separating the infected from the healthy. Exclusion of those infected with a contagious disease is the surest way to break the course of transmission.

By the 1940s, new wonder-drug antibiotics made headway against *Mycobacterium tuberculosis,* but soon doctors noticed their therapies were not working as well as before. Antibiotic-resistant mutants were emerging. In the following two decades, a host of alternate antibiotics were introduced to combat the new strains. When the drugs' effectiveness disappeared, doctors began using the antibiotics in combinations to continue treating their TB patients. Few people in medicine recognized that this decline in effectiveness was signaling the advance of resistant strains. If a handful of doctors did voice concern, their worries were calmed as new TB infections in the United States decreased over the next thirty years.

Worldwide, the average yearly incidence in the last ten years has increased by less than 1 percent. This slow increase may suggest that TB is not a threat. Unfortunately, the disease is a persistent menace in many countries, and it is expected to steadily increase. WHO estimates that up to one-third of the planet's population is infected with TB. Two to three million people die from it each year.

Today doctors juggle a combination of four antibiotics to eliminate TB from the lungs. They cross their fingers that this treatment will reduce the chances of promoting new resistant strains. Even without the risk of a resistant microbe complicating a patient's recovery, treating TB has never been an easy process. TB chemotherapy takes months. The risk of resistance grows if patients do not follow the treatment for its entire prescribed course. Even with proper therapies, *Mycobacterium tuberculosis* lives as a parasite in the lungs and is difficult to defeat. Furthermore, only two or three cells are needed to start a repeat infection.

Some of the reasons for TB's resurgence are not easy subjects to discuss. Immigration and changing demographics are thought to influence the disease's patterns in North America, and its reemergence in the United States is expected to become a major health concern within the next five years. Unlike smallpox eradication, initiatives for arresting the global spread of TB have been slow to develop.

The major reasons for the reemergence of TB are as follows:

- Global economic conditions. The highest incidences are seen in areas with the lowest gross national product. Current treatments are long term and expensive and therefore create a barrier to successful control, especially in developing countries.
- HIV infection. TB is an opportunistic disease associated with HIV infection. WHO estimates that of HIV-positive individuals, one in ten will develop TB. The worldwide increase in HIV infection is leading the way toward increased TB infection.
- Multidrug resistance. TB strains that are resistant to just one or two antibiotic treatments may now outnumber susceptible strains. Four-drug regimens (isoniazid, rifampin, pyrazinamde, and ethambutol) are strongly recommended to outmaneuver resistant strains. When patients do not follow their therapy to its conclusion, they open the door to recurring infections, which may then take up to two years to treat.
- Immigration. Immigration from countries with a high TB incidence helps the disease return to previously "safe" countries. In the United States, nearly half the new cases are in immigrants. Surveillance of the disease may be complicated in areas with high numbers of immigrants and where cultural and language barriers may exist.
- Complacence. When the rate of new cases of a disease begins

to slow or decline, the public health community might conclude that it is no longer a threat to the general population. Less emphasis is then placed on research and education.

• Urbanization. Migration to cities has been a recurring theme when studying mankind's relationship with infectious disease. Today, the TB threat is greatest in places where crowding and close personal contact is commonplace: prisons, long-term health facilities, nursing homes, and homeless shelters. With increased urbanization, infectious disease will have increased opportunity to move through each population.

In 1900, 15 percent of all people lived in cities. In 1950 the figure was at 30 percent. Presently, about half the earth's population lives in cities, and in twenty-five years the percentage is expected to be 65 percent, or 5.2 billion people. Opportunities for *Mycobacterium tuberculosis* are looking brighter every day.

An Emerging Disease

An emerging disease is one about which nothing was known prior to its entering a population (the AIDS virus in the 1970s United States), or one for which the infectious agent is known to exist only in animals (Table 7.1).

The list of new infections from bacteria, viruses, fungi, and protozoa is expanding. Towns and communities are encroaching into previously undeveloped land, thus making conditions favorable for new pathogens to make an animal-to-human jump. In addition, techniques for identifying microbes have become more accurate. Global warming is also expected to influence disease by affecting the numbers of insect vectors and animal reservoirs that carry pathogens.

One well-known microbe that seems to consistently invent a new wrinkle in its virulence is *E. coli.* For decades *E. coli* has been a

Prevalent Emerging Diseases Worldwide

Emerging Infectious Diseases	Reemerging Infectious Diseases
West Nile virus (West Nile encephalitis)	*Mycobacterium tuberculosis* (Antibiotic-resistant TB, XDR-TB)
Nipah virus (Encephalitis)	*Staphylococcus aureus* (MRSA)
Hendra virus (Encephalitis-like symptoms)	*Streptococcus pneumoniae* Antibiotic-resistant pneumonia)
Prions (Creutzfeld-Jacob Disease)	*Corynebacterium diptheriae* (Diptheria)
E. coli O157, type O124 (Foodborne illness)	Dengue virus (Dengue fever)
Vibrio cholerae O139 (Cholera)	Yellow fever virus
Hepatitis E virus (Hepatitis)	*Yersinia pestis* (Bubonic plague)
	Influenza virus reemerges yearly

Table 7.1. New diseases will appear in the future.

cooperative microbe in laboratory experiments. Over time new serotypes were discovered, and microbiologists began grouping *E. coli* strains according to the type of disease they caused. In 1982, the O157:H7 serotype was identified as the cause of a number of food-borne outbreaks and reemerged dramatically in 1993 when an outbreak in the western states affected over 700 people and resulted in the death of four children from severe kidney failure.

Lately, a new type of O157 has emerged, adding another layer of complexity to the ways *E. coli* can create havoc. A subtype of *E. coli* O157 called EXHX01.0124, or 0124 for short, has burst onto the scene. This variety of *E. coli* is similar to O157 in that its harm comes from a neurotoxin released by the cell. Though the CDC had tracked a small number of foodborne illnesses to O124 since 1998, it made big headlines in the fall of 2006 when outbreaks in twenty-six states and one case in Canada were tentatively traced to fresh spinach grown on commercial farms in California. Another outbreak later in the year was associated with California-grown scallions, the occurrences this time in New Jersey and Long Island, New York.

When a new pathogen such as O124 emerges, there are few clues for public health authorities to use in advising a worried public. It may take months to determine the routes a specific strain follows when entering the food distribution network. The O157 strain had been linked years earlier to outbreaks from undercooked hamburgers, and the public eventually learned the dangers of undercooking hamburger meat.

The O124 strain threw a curveball at consumers because most of them may have regarded fresh fruits and vegetables as the safest and healthiest part of their diet. Considering the details of a previous *E. coli* outbreak, perhaps growers should not have been caught off guard by the O124 incident. In 1996, a maker of fruit juices distributed *E. coli* O157-contaminated products to several western states and Canada. At least seventy persons became sick with gastrointestinal symptoms; the most seriously affected were children. Among the children, one girl in Colorado died. After detective work by public health microbiologists, the trail traced back to O157-contaminated apples that went into the unpasteurized juices.

How could an intestinal microbe like *E. coli* O157 wind up on an apple growing on a tree? How, too, did the later outbreak come about in which the O124 version of O157 contaminated fresh spinach, scallions, and other suspect vegetables? All *E. coli* bacteria come from animal intestines and so are now known to be spread among animal carcasses during meat processing. Cattle, sheep, and other animals are sources of *E. coli* on vegetable crops in the field. Runoff from feedlots, periodic flooding of vegetable fields with water drained from feedlot farms, manure fertilizer, and even wild animals passing through the fields are all probable points of contamination. The apples used for making the tainted juice were not the unpicked ones, but those apples that had fallen to the ground, greatly increasing their chances of contamination with feces.

What Makes a Disease Emerge or Reemerge?
Emergence and reemergence of infectious microbes often happen because of changes in well-established patterns. Foods are centrally processed in more and more countries. They are distributed to larger geographic areas than they were a few decades ago. The economics of food supply have become highly efficient. So, too, has become the spread of foodborne pathogens. Health choices played a part in the outbreaks from fruit juices. A desire for "healthier" foods may have prompted the idea of unpasteurized products to meet consumers' wishes. By removing one facet of the food safety process, the juice manufacturer actually created a more dangerous situation, and delivered to the world a "new" pathogen. In 1996, *E. coli* O157 was defined as an emerging pathogen. Today, it is a recognized contaminant. The 0124 subtype might be following in O157's footsteps.

Emerging diseases are not only frustrating for doctors and microbiologists but also alarming for the affected community, which feels helpless and vulnerable. People demand answers from the health community and politicians, and there may be no answers until microbiologists uncover the details of the pathogen's transmission and its virulence.

Certain microbiologists confine their work to emergence/reemergence. Some focus on one disease. In broad terms, recent emerging or reemerging pathogens tend to be viruses easily transmitted between persons, with the emergence most likely in geographic regions undergoing changes in population (migrations, aging) or ecological changes (construction in previously undisturbed areas).

A population is at risk only when the pathogen finds a significant number of susceptible hosts, has sufficient virulence, and is offered opportunities to infect. Once a new pathogen begins its relentless march into our lives we can barely remember the days when it didn't have a name. The uncommon pathogen that once seemed distant and innocuous becomes a microbe that lives with you.

Germs at Play

One reason an infectious disease might emerge or an old one reemerge has to do with changes in normal routine. Altering habits and regular schedules can give a sliver of opportunity to an infectious agent. A change in your regular daily practices serves two purposes for inviting infection: (1) it potentially puts you in proximity to a new pathogen, and (2) it increases your susceptibility as a host.

"Stress" isn't standard medical jargon. It found a place in medical language in 1936 when it was coined by the Austro-Hungarian physiologist Hans Selye to mean "the non-specific response of the body to any demand for change." Any definition that uses the words nonspecific invites deliberation. Before long, scientists from varied disciplines wrangled over the meaning of stress. Selye recognized the difficulties and eventually introduced the term "stressor" to differentiate between the things that cause discomfort and the discomfort itself. (For his efforts, Selye earned the dubious sobriquet "The Father of Stress.")

Today everyone seems to know about stress even if they find it hard to describe. References to stressful lifestyles and stress-related terms are in our lexicon: "stressed out," "burned out," "wired," "chill," "timeout," and "downtime." Magazines publish articles on the top ten Stressors. It's questionable if these lists are clinically proven considering the medical community itself has not identified all the factors that create stress in humans and other animals. But psychologists recognize a long list of key stressors in society, and health risk analysts are able to draw connections between some of them and the chances for infection. One stressor, believe it or not, is vacation.

Vacations and leisure activities offer new avenues for infection. When traveling on vacation, you may be short on sleep and good nutrition, thereby stressing your immune system. Vacation travel and destinations may put you in crowds where germs spread easily— a cruise ship, an amusement park, air terminals, a street festival. Hygiene often goes on vacation when people do. When lax hygiene coincides with heightened host susceptibility and increased opportunities for infection by pathogens, the result is no day at the beach.

Oceans

The beach is almost everyone's first choice as a vacation spot. The microbes in oceans and lakes make up perhaps the most fascinating aspect in all environmental microbiology. Recreational waters (oceans, seas, bays, lakes, rivers, etc.) contain a mixture of bacteria, protozoa, algae, diatoms, and microscopic parasites, to name just a few organisms. As you might expect, some are pathogens.

Salty ocean water does not kill germs as many people believe. Humans have been swimming in a diverse mixture of marine microbes through the millennia. There are estimated to be at least 20,000 different kinds of bacteria in a liter of seawater. A mouthful of seawater taken in by a swimmer contains about 1,000 types of bacteria. One cc can contain over 100,000 microbial cells.

The ocean's salt concentration is 3.5 percent, a level not harmful to most microbes. The ocean's resiliency in accepting civilization's wastes has long lulled society into the carefree belief that ocean water would put our refuse out of sight forever. Scientists today agree that the oceans' function as a toilet bowl for humanity must end.

Humans produce about 0.4 to 2.0 pounds (0.2–0.9 kg) of urine and feces daily. There are over 6.5 billion people on earth. The majority live near coasts and major tributaries. Aboveground runoff and overflowing sewers carry wastes downhill to open waters. Remember also the enormous amounts of domestic and wild animal feces that runs off from farms and woodlands to rivers, bays, and oceans. The world is not a perfect place. Sooner or later bad germs will show up in nice places.

Beach waters are tested regularly for *E. coli* and enterococci, which are a variety of round bacteria that live in the intestines. Most disease-causing microbes shed by animals are in low levels in nature, and the methods to detect them are therefore expensive. *E. coli* and enterococci behave the same as these would-be pathogens, so water quality microbiologists use their presence as an indication of potential fecal contamination from humans, animals, or birds. For that reason *E. coli* and enterococci are referred to as **indicator bacteria.** Indicator bacteria in beach waters do not prove that human wastes are present but are a sign that contamination may have occurred and pathogens may be there. The EPA sets the maximum bacterial levels allowed at marine beaches at no more than 35 enterococci per 100 milliliters (about 3 and 1/3 ounces) of water.

Microorganism levels are highest at beaches near outflow pipes clearing runoff from neighborhood streets and yards. High levels are also found at some of the most family-friendly beaches, those calm and shallow areas often protected by jetties. Calm beaches are not vigorously churned. At these places gentle waves disperse less sand and particles to which bacteria tend to attach, so contamination remains locally concentrated. Late spring–early summer and late summer–early fall have the highest microbial numbers. Phytoplankton levels are also highest during this time, suggesting the overall biological activity of the ocean is at its height.

Increased river flows and heavy rainfalls dilute ocean salinity (salt levels). With the right combination of water temperature and nutrients, red algae, a type of plankton, grow rapidly to huge amounts. When this happens, it's called an algal bloom. The algae float near the water's surface and soak in sunlight while taking advantage of the lower salt levels. High concentrations give the water a reddish color, so the phenomenon is called **red tide.** Red tide turns water at the beach smelly and slimy. Red algae also produce a neurotoxin that causes severe irritations to the eyes and skin, and may cause fever and vomiting. Humans can get sick from eating seafood that has ingested red algae. Fish, too, are susceptible to the toxin. In recent years different types of algal blooms all over the world have caused massive die-offs of fish, events known as fishkills.

Algal blooms are made up of different types of algae, so the term harmful algae blooms (HAB) is starting to replace "red tide." Other neurotoxins are produced by a variety of unicellular plankton called dinoflagellates. Domoic acid is a toxin produced by a diatom, concentrated in mussels, that causes fishkills and marine mammal deaths. There have been cases of domoic acid poisoning in humans that have eaten contaminated shellfish and possibly shrimp or crabs.

Red tides are a natural reaction to cyclical environmental conditions. They have been known for centuries, as evidenced by biblical references: "And all the water that was in the Nile was turned to blood. And the fish that were in the Nile died, and the Nile became foul" (Exod. 7:20–21). Modern pollution output from industrialized areas and farmlands also contributes to algal blooms. The high amount of organic compounds found in sewage, industrial wastes, and fertilizer and manure from farming operations are some examples. Once the bloom grows to such an extent that the algae deplete oxygen from the surrounding waters, marine life suffocates.

Boaters can be a significant source of food and fecal contamination added to waters that are near bathing beaches. The EPA has established NDZs, No-Discharge Zones, to prohibit sewage and garbage dumping near recreational areas. Enforcement is the problem. Like many other regulations set in place to protect the environment, they only work as well as the level of responsibility of boaters, hikers, and campers.

Indicator bacteria give a partial picture of the microbes in the swimming waters. Indicators don't predict *Giardia* and *Cryptosporidium* levels, which are high near municipal outflow pipes, and they give no indication of virus levels. Marine samples from beaches near large cities may contain up to a million viruses per milliliter (1 cc). Many of the unexplained water-associated outbreaks are thought to be caused by viruses. Viruses or virus groups considered to be health risks in recreational waters are hepatitis A and E, caliciviruses, rotavirus, adenoviruses, Norwalk viruses (noroviruses), and astroviruses. The hepatitis viruses A and E are found in all waters and are a problem in countries where poor sanitation consistently pollutes inland and marine bodies of water.

Swimmer's ear (acute otitis externa) is caused by a bacterial secondary infection after the ear canal becomes irritated by ocean water or freshwater. Ear wax, or cerumen, provides a physical protection to the ear canal and lowers the pH to around 5, which inhibits some non-native bacteria. Frequent swimming can disrupt the wax layer; the skin's protection becomes compromised and the pH rises to 7. The ear's normal Gram-positive staphylococci are then overrun by Gram-negative bacteria, especially *Pseudomonas*. *Aspergillus* mold and *Candida* yeast also appear in cases of otitis externa. The ensuing inflammation caused by these opportunistic microbes creates a moist and nutrient-rich environment that further promotes microbial growth.

Itching that comes from swimming should not be scratched. A finger or the aggressive use of cotton swabs or other probes only exacerbates the problem. Pain follows as the inflammation intensifies. Topical antibiotics are usually adequate to reduce *Pseudomonas*, and the other bacteria common in case of swimmer's ear, *Proteus* and *Staphylococcus*. The antibiotic polymyxin is effective against

The Hepatitis Viruses

Hepatitis is inflammation of the liver and can lead to jaundice. **Hepatitis A** (HVA) and **hepatitis E** (HVE) are associated with contaminated drinking and recreational waters. Both are transmitted by the fecal-oral route, but HVA is also a foodborne pathogen and a sexually transmitted pathogen, and is the most prevalent type of global viral hepatitis cases. As few as ten HVA particles may be enough to start an infection. HVE is immunologically different from HVA and rarer. All types of hepatitis cause the following symptoms to greater or lesser degree: fever, malaise, nausea, loss of appetite, and abdominal pain.

Hepatitis B (HVB) infection is almost as frequent as HVA, but HVB is not found in water. It is transmitted through blood transfer, either in transfusions or sexually. **Hepatitis D** (HVD) is transmitted sexually and through blood-to-blood transfer. It gives the same symptoms as other hepatitis viruses, but needs a preexisting HVB infection in order to infect. Over 4 million Americans are infected with **Hepatitis C** (HVC), and 80 percent of those infected do not know they have it. Sexual transmission of HVC is possible, but most transmission follows a blood-to-blood route. HVC is the leading reason for liver transplants. Hepatitis B, C, and D are chronic diseases.

There is no vaccine for HCV and treatment may be only 50 percent effective. There are effective vaccines against HVA and HVB. HVD is prevented through HVB vaccination.

Pseudomonas; neomycin works against *Proteus* and *Staphylococcus;* Nystatin works against *Candida.* Frequent swimmers prevent otitis externa by using fitted earplugs or by applying eardrop solutions to keep the pH from rising. A do-it-yourself formula is prepared by mixing white vinegar with 70 percent rubbing alcohol.

Lakes, Streams, and Other Freshwaters

Lakes and ponds, rivers and streams are all favorite summertime places to cool off and enjoy the outdoors. Unfortunately they also have their share of waterborne microbial menaces. The variety of infectious agents in freshwater is sometimes greater than that found in saltwater. Pathogens in various freshwater sites have been responsible for infections of the skin, eyes, urinary tract, respiratory system, and central nervous system. Fresh surface waters are the first collection points for aboveground runoff after rains. Untreated groundwaters, such as contaminated wells, are also pathogen sources. In extreme cases of infrastructure failure, sewage may leak into lakes and streams before reaching the sewage treatment plant.

Waterborne microbes use various routes to enter the body: inhalation, through intact or injured skin, or infiltration of the mucous linings of the respiratory or digestive tracts. The chances of infection from freshwaters may be slightly higher because people tend to swim in freshwaters for more minutes at a time than they do in chilly ocean waters where the salt and exertion needed to battle the waves act to shorten the swim. This is admittedly a broad generalization; surf riding, and some snorkeling and scuba diving can last several minutes to an hour in the ocean.

The rules for swimming in freshwater are the same as for oceans and bays. Hand-to-face touching should not take place. Shower after a swim. Finally, never swim with an open cut or a known infection. (Waterproof adhesive bandages do not give cuts adequate protection against infection.)

Freshwater may contain the same viruses as found in saltwaters plus *Giardia* and *Cryptosporidium* cysts, *Entamoeba histolytica* protozoa, and the bacteria *Salmonella, Pseudomonas, Shigella, Yersinia,* and *Vibrio. Salmonella typhi* is the cause of typhoid fever. *Vibrio cholerae* causes cholera, a disease that is reemerging worldwide.

Pseudomonas bacteria do not do well in saltwater, but are ubiquitous in freshwaters. In fact, *Pseudomonas* is the predominant genus found in drinking water as well as a member of biofilm communities. This microbe has been implicated in the following infections associated with exposure to freshwater: dermatitis, folliculitis, swimmer's ear, ocular keratitis, urinary tract infection, and pneumonia.

In freshwaters, an infectious microbe like *Pseudomonas* may be native or it may be one of many contaminants, that is, *Giardia, Salmonella,* hepatitis A. Swallowing water during a swim increases the odds for getting a gastrointestinal ailment. But swimming with a preexisting infection of the skin, eyes, or lungs opens the door for secondary infections from the long list of waterborne microbes.

The EPA sets the maximum allowed freshwater levels of indicator bacteria at 126 *E. coli* per 100 milliliters and 33 enterococci per 100 milliliters. *E. coli* is a slightly better indicator of fecal contamination in freshwaters; enterococci are the preferred indicators for both fresh and marine waters.

Swimming Pools, Hot Tubs, and Steam Rooms

Warm water, constant aeration, and nutrients supplied by bathers make a perfect recipe for bacterial growth. Expect whirlpools, spas, and hot tubs to contain more microbes than pools because of warmer temperatures and vigorous agitation. A higher number of users in a smaller volume of water also put more organic matter into spa-type water, and this enhances microbial growth.

Most swimming pool owners understand the need for chlorination so pools tend to be monitored better than backyard hot tubs.

Pool microbiology is helped, too, by cooler temperatures and less turbulence. Chlorine levels remain more stable and for a longer time than in turbulent water. For this reason, spa-type waters should be chlorinated at slightly higher levels than pools. Most pools are maintained at 1 to 3 ppm of free available chlorine. Whirlpools and hot tubs should be maintained at 1 to 5 ppm. All chlorination must be at pH values in the range of 7.2 to 7.6. At pH levels lower than this range, chlorine becomes corrosive. At higher values, the effectiveness of chlorine as a disinfectant drops dramatically.

Hot tub temperatures are not too hot for microbes. Most hot tubs are set to about 100 degrees F (37.7 degrees C) and never higher than 104 degrees F (40 C). These temperatures are within the optimal range for the bacteria that can cause infections. Bacteria and viruses can withstand short bursts of temperatures much higher than those that are safe for human skin. For example, to kill the waterborne hepatitis viruses A and E, water must be boiled for one minute.

Spas provide environments where *Pseudomonas* and other common water microbes form biofilms. Because high numbers of microbes accumulate in biofilms, their presence on the hot tub or whirlpool wall will greatly elevate the total amount of microbes in the spa water. Biofilms are very resistant to chlorine disinfection.

Strict attention should always be paid to chlorine levels and pH in heated recreational waters. In warm conditions that are part of the enjoyment of hot tubs and steam rooms, the skin's pores dilate. This dilation may help microbes like *Pseudomonas* or fecal microbes to enter into deeper layers of the skin. Folliculitis and dermatitis from *Pseudomonas* infection are well documented. A more serious condition known as "hot-foot" syndrome is associated with abrasions on the soles of the feet when exposed to rough pool floors. In this situation, the microbe enters deep layers of the skin and sets up painful microabscesses and lesions.

People who use steam rooms have the advantage of avoiding the curious admixture of others' germs floating all around them. But steam rooms are still moist environments with heavy use by many strangers. The surfaces are touched constantly and clothing is not worn. All these factors promote the transfer of bacteria and yeast from one person to the next. Towels should always be used to avoid sitting directly on benches, and shower sandals should be worn. Sandals should also be worn in the locker rooms and while showering.

Each state has health codes that describe the requirements for public swimming pools and recreational waters. The codes include actions that operators must take in the event of fecal contamination. The state health codes can be found online or by contacting the National Swimming Pool Foundation.

Gyms and Health Clubs

Avoiding microbes in the gym involves the same commonsense procedures you would use anywhere else. The gym offers a lot of worst-case scenarios for spreading germs. In athletic clubs there are numerous vehicles for transmission (equipment, towels, benches, etc.): moisture from sweat, many strangers handling shared surfaces, opportunities for scrapes and abrasions, and confined areas such as shower rooms, whirlpools, steam rooms, and saunas. Sweat itself does not contain microbes, but it acts as a good moisture-laden vehicle for germ transfer from place to place. (Sweat can pick up bacteria from the skin as it exits the sweat gland.) Sweat contains proteins and salts, which provide nutrients to bacteria and therefore help them hang around longer on inanimate objects. If pathogen numbers are in the range of the microbe's infective dose, then vehicle transmission of infection in the gym is plausible.

Perhaps the gym's biggest hazard is distraction. Most people are not thinking of hygiene during a spirited workout. In this way, the gym is like your kitchen or your office at work. Losing sight of good hygiene anywhere is the surest way to become infected with other people's germs.

Microbial studies on gyms have identified many familiar microbes: staphylococci, streptococci, and Gram-negative bacteria on all the equipment; *E. coli* and other fecal organisms almost everywhere; *Candida* yeast on equipment seats; and athlete's foot fungus in showers and locker-room floors. Increasingly, health clubs are becoming a site where **MRSA,** methicillin-resistant *Staphylococcus aureus,* is being contracted. Because this gym-associated "superbug" is not connected to its usual haunt, hospitals, it is sometimes referred to as CA-MRSA, community-acquired MRSA. At present, MRSA is known to be resistant to fifteen to twenty different antibiotics. In a 2005 British study, at least one hundred men and women had contracted the bug in gyms and health clubs.

Health officials in England and Wales tracked deaths due to MRSA over a twelve-year period. In 1993, they recorded about 400 deaths with almost all attributed to regular (non-MRSA) *Staphylococcus aureus.* By 1999, there were 1,000 deaths, and the MRSA variety made up 50 percent. In 2004, over 70 percent of the 1,700 mortalities were due to MRSA with less than 30 percent from regular *Staphylococcus aureus.*

MRSA infections range from localized skin infections to septicemia and shock. As with other infections, the elderly, hospitalized, or immunocompromised are at the greatest risk of serious infection and death.

The good news about gym germs is that they are the same ones normally found on everyone's bodies. The bad news about gym germs is that the high frequency of touching surfaces and high numbers of people in close proximity increase the chances for infectious pathogens to spread. The gym is no different from other crowded places. Extra care should be taken to stay healthy and safe whenever you visit an athletic club. Be particularly vigilant if it's cold and flu season or if a person on the treadmill next to you shows signs of being sick. The familiar warnings are coughing, sniffling, sneezing, deep coughs with phlegm, or obvious congestion. As always, do not touch hands to your ears, eyes, or mouth.

Follow these safeguards to prevent germ transmission in gyms:

• Wash hands before and after exercise.

• Wipe down equipment with the disinfectant sprays or disinfectant wipes provided by the gym, before and after exercise. If the gym doesn't provide wipes, ask for them.

• Bring your own towels from home and do not use them to wipe down the equipment. Use them for your own body sweat.

• Avoid touching any surface directly with any part of your body other than your hands. Sit on towels when using locker-room benches, saunas, steam rooms, and exercise equipment. Wear sandals.

• Be mindful of the many places that members share each day in a gym: equipment seats and handles, benches, yoga mats, gymnastics and wrestling mats, weights and medicine balls, plates and dumbbells, exercise balls, racket handles, basketballs, jump-rope grips, and resistance bands.

• The restroom at the gym is a public restroom. Treat it as such with all the normal precautions.

• Never reuse your towel. Wash it immediately or put it in the hamper at home, and then take a clean one on your next visit to the gym.

• Treat athletic clubs like restaurants. Ask for a tour before signing up for membership. As the guide points out state-of-the-art elliptical trainers, look for telltale signs of poor building hygiene: little or no ventilation, dusty and grimy ventilation ducts, dust and dirt on the floor and in corners, poorly maintained bathrooms, dim lighting, and employees wearing sloppy or dirty uniforms, or otherwise not following good personal hygiene practices.

The routes of germ transmission in gyms are concentrated in one crowded building, a concern avoided during outdoors activities. Participants in outdoor athletics and team sports might, however, have slightly increased chances of scratches and scrapes, loosened teeth, and even assaults to the eyes, ears, and nose. Outside or inside, the modes of transmission in sporting events are direct person-to-person contact, indirect transmission through shared equipment, and indirect transmission in droplets from sneezing, coughing, or the huffing and puffing that comes with exertion (Figure 7.1), all with greater frequency than in more pedestrian activities.

Figure 7.1. These fellows understood the hazards of germ transmission during athletics (1918 flu pandemic). *Copyright CORBIS.*

Cruise Ships and Hotels

Upscale ships and hotels expose vacationers to the places where germs are easily transmitted: beaches, pools, spas, health clubs. Cruise ships and hotels are packed with a semiconfined group of people repeatedly sharing the same objects and surfaces. Guests may use the same on-site restaurants and water sources exclusively for a week or more. Outbreaks of contagious illnesses and foodborne pathogens on cruise ships seem to make consistent headlines. Unlike the world of ever-moving air travelers, tracing an outbreak to a single cruise is relatively easy. Despite the negative publicity received by cruises and favorable conditions for transmitting onboard pathogens, most trips are infection free.

One risk on ships and in hotels is *Legionella*. This bacterium received its name in 1976 when it was cited as the cause of pneumonia that broke out among American Legion conventioneers at a Philadelphia hotel. Of the 182 infected, twenty-nine died. The species *Legionella pneumophila* soon became famous, and a new disease emerged: Legionnaires' disease. Since then, new methods are used to identify *Legionella* when there are potential outbreaks at conferences, hotels, and on cruise ships.

Legionella's main mode of transmission is water droplets or fluid vapors inhaled into the lungs. Elderly smokers are at the highest risk for contracting Legionnaires' disease. Air conditioning units and cooling towers have been implicated, but vapors from water devices such as humidifiers and whirlpools are now thought to be the main mechanism in transmission. In very rare instances, inhaled water from hot springs, decorative fountains, and water-containing dental devices have spread the pathogen.

Today monitoring for *Legionella* has helped control outbreaks. Copper pipes inhibit microbial growth while plastic PVC pipes help most bacteria attach to inner surfaces and grow. Hot water systems maintained above 130 degrees F (54.4 C) also inhibit *Legionella*. In addition, water treatment specialists now recommend specific chemical biocides for preventing growth in nonpotable (nondrinkable) water devices and distributors. Chlorine, chlorine dioxide, ozone, and DBNPA (dibromo-nitrilopropionamide) powder are most effective for removing *Legionella* from non-potable water.

Foodborne illness outbreaks occur on ships for the same reasons they do in restaurants. The rules for good food handling and proper personal hygiene are no different onboard a boat. There is no evidence that the incidence of outbreaks on ships or in hotels is greater than in other food-preparation establishments.

Many cruise lines or individual cruise ships are inspected by the CDC under the Vessel Sanitation Program, which was established in the 1970s after a rash of outbreaks on ships. VSP officials conduct sanitation inspections, disease surveillance, and construction plans for new ships, and provide hygiene training for shipboard staff. Among the items inspected are water supply, spas and pools, food preparation, employee hygiene practices, and general ship cleanliness. Individual ships and lines that have volunteered for the VSP receive scores postinspection. The inspection results may be accessed by searching the Vessel Sanitation Program on the CDC's Web site.

Summary

You can't take a vacation from germs. Good hygiene and awareness of the microorganisms around you help prevent infections that could turn a fun getaway into a nightmare. Although MSRA has been recognized for decades, its incidence of infection is increasing; it is an emerging agent. Changes in people's behavior and demographics contribute to how and why diseases emerge or reemerge. Emerging microbes seem far away from our daily lives and nonthreatening. But every pathogen known today was at one time a new and unsuspected danger. Knowing the basics of germ spread and its prevention help keep newly discovered microbes out of our communities and the harmful ones out of your life.

8

On the Horizon

If we consider all these data it is well-nigh impossible
not to be impressed by the enormous diversity in microbial
metabolism.

—Albert J. Kluyver

After seven chapters of bacteria, viruses, mildew, and mold, it may now be easier to spot the microbial and antimicrobial activities found in your home. Food and water are two of the home's primary entry points for both harmless and potentially dangerous microbes. But many foods contain an arsenal of chemical additives to combat microbial contamination. Other foods use their natural characteristics to prevent spoilage from microbes. There are methods at your disposal, too, for reducing the risk of ingesting a food pathogen. Under the kitchen sink and in the medicine cabinet, most homes carry another cache of antimicrobial products to ward off microscopic interlopers.

The good microbes are all around, and they outnumber pathogens by several thousand to one. Of the millions of species of known and unknown microbes on earth, pathogens make up a minuscule portion. Among pathogens, about 540 are bacteria, about 320 are fungi, 200 or so are viruses, and less than 60 protozoa cause illness. The silent majority of good microbes have truly lost to the pathogens when it comes to the publicity they get.

It would be virtually impossible to plan a meal that doesn't include food or drinks made by microbes. Microbial biochemical activities often create a hostile site for invading spoilage microbes or pathogenic ones. Your skin bacteria also silently fight off a daily array of pests that can cause disease if left unchecked. Millions of intestinal bacteria help digest food and their sheer biomass also provides you with proteins and vitamins. If a native microbe finds an opportunity to invade, the immune system launches an exquisitely cooperative team of defensive specialists, cells that destroy any pathogen or nonpathogen that enters the bloodstream.

No system is perfect. Infections happen. Sometimes a grave condition follows. Often the process is aided by forgetting about good hygiene. Maybe you have remembered proper hygiene principles, but the person sitting next to you with the flu has ignored them. Indeed, you are often at the mercy of things out of your control. You may have little defense against a drug-resistant species inhabiting your home or local hospital. The risks increase if you are elderly, or worn down by stress, or have a weak immune system.

The dangers of microbes remain all around you, but so do the benefits. You are now beginning to see the world through a microbiologist's eyes. Look closer. There's even more microbiology in your life. As biology and technology advances, a new generation of microbial benefits is just around the corner.

The Hidden Microbes at Home
Some microbial products in the home may not be as easy to spot as a glass of wine or a slice of cheese. An example is xanthan gum. This large molecule produced by *Xanthomonas campestris* bacteria works as a thickener in personal care products and canned sauces, bottled salad dressings, and ice cream. In the production of wheat-free products, xanthan gum replaces gluten for adding volume and viscosity to bread. In agriculture, as opposed to manufacturing, *Xanthomonas* is less than desirable. It is a plant pathogen that causes black rot disease in vegetables such as broccoli, brussels sprouts, cabbage, kale, and cauliflower.

What does contact lens cleaner have in common with a clogged kitchen drain? Products for cleaning lenses and eating through drain clogs may contain the enzyme subtilisin. Subtilisin was first discovered as a product of *Bacillus subtilis,* bacteria you pick up with every clump of garden soil. Enzyme companies have harnessed *Bacillus subtilis* and other microbes to produce copious amounts of it for various commercial uses. Subtilisin is used in laundry detergents, stain removers, and dishwasher tablets to degrade proteins. When poured into clogged drains, it gradually breaks down a variety of proteins from meat, fish, cheese, and eggs. Other microbially produced enzymes in drain openers eat through the clog's fats and oily components. Still more digest starchy materials. Septic tank treatments similarly use subtilisin and other enzymes to decompose wastes.

Hair is one thing in bathroom drain clogs that is almost impossible to degrade. Human hair is made structurally strong by a group of proteins known as keratins. The amount of keratins in hair varies by race, but all hair has in common a strong resistance to enzymatic breakdown. Very few microbes degrade hair, or if they do, they cannot degrade it fast enough to get your bathroom drain flowing again within minutes.

Your wardrobe is in large part the recipient of microbial reactions. Before leather is shaped into shoes or jackets it undergoes rigorous treatment. Hair and oils are removed. The leather is treated to become pliable, and then stretched to the desired thickness before being tanned. The leather industry uses lime and other strong chemicals on untreated hides, but enzymes from bacteria and fungi are beginning to replace some of these chemicals. In addition to subtilisin from *Bacillus,* other microbes that produce protease used in leather treatment are *Streptomyces* and *Rhizopus* bacteria and *Aspergillus* and *Penicillium* molds. *Aspergillus* lipase, an enzyme that breaks down fats and oils, is also used in processing leather.

Fabrics, too, receive enzymatic treatment. In making fabric, threads are coated with a protective material in order to reduce damage during weaving. Before the woven fabric is dyed, the coating must be removed. The microbial amylase enzyme from *Bacillus sub-tilis* and a number of molds digest the fabric's starch components. Cellulase enzyme polishes the fabric, meaning it helps remove small fuzzy balls. Microbially produced cellulase is also used for stone-washing denim. *Trichoderma* yeast is a good source of cellulase for these purposes.

High-Tech Microbes

There is a vast resource of microbes that will be engaged in coming years for new products or for the improvement of existing products. One example is the fungus *Aureobasidium pullulans,* which makes pullulan. This polymer (a large molecule made of a many repeating subunits, such as sugars) may have potential as a component in degradable food wrap. Other polymers excreted by the bacteria *Pseudomonas, Alcaligenes,* and *Syntrophomonas* are gradually being put into use in biodegradable plastics.

Manufacturers take advantage of special talents not found in all microbes. For instance, the enzymes in laundry detergents must work in hot water. The microbes that produce these enzymes are therefore those that thrive at high temperatures. Conversely, cold-water detergents use enzymes from bacteria that are accustomed to living in chilly environments such as soil, cold locales such as deep water, or the frozen environment within ice.

Some bacteria flourish in environments where few other living things exist. The advantage to the bacteria is obvious; there is no competition for nutrients or niche in a place where nothing else can grow. The advantage to people is that we can utilize the skills of these specialized bacteria.

The Extremophiles

Extremophiles are bacteria, fungi, or algae possessing attributes that allow them to live in extreme or harsh conditions that few microbes, plants, or animals can withstand. In recent years, environmental microbiologists have put a number of extremophiles to work in biotechnology and hazardous waste cleanup.

Extreme Temperature Bacteria

Temperature is one feature that defines an environment. Bacteria have a favorite or optimal temperature range in which their biology is at its most efficient. The bacteria on your body and in the household prefer the same temperatures in which humans are comfortable. These bacteria are called **mesophiles.** For instance, *Staphylococcus aureus* on skin prefers 60 to 120 degrees F (15 to 50 degrees C) for growth, but can survive at 145 degrees F (63 degrees C), or lower to 40 degrees F (4 degrees C). Outside their optimum, mesophiles grow very slowly or not at all. This explains the need for cooking foods to at least 140 degrees F (60 C) and keeping your refrigerator at 35 to 40 degrees F (2 to 4 C). Bacterial enzyme systems come to a virtual halt when frozen. The bacteria begin growing again once they thaw. Repeated freezing-thawing cycles are more destructive than a single temperature change. During these cycles small ice crystals form inside the cells every time they freeze. The crystals eventually cause irreparable damage. A single heating is more destructive than a single freezing. Temperatures above 145 degrees F are lethal to mesophiles.

Thermophiles are microbes that live at very hot temperatures, up to 165 degrees F (74 C) or thereabouts. Hyperthermophiles survive at even higher temperatures like the conditions at hot springs and geysers, thermal pools, and volcanic eruptions. A group of microbes called *epsilon*-proteobacteria are hyperthermophiles that have been discovered living at hydrothermal vents a mile below the surface of the Pacific and Atlantic Oceans. The temperatures at these vents exceed 480 degrees F (250 C), similar to those at the earth's surface billions of years ago. *Epsilon*-proteobacteria are not alone down there. Bizarre tube worms and crustaceans have been also been found near submarine hydrothermal vents.

Psychrophiles are at the cold end of the thermometer. Many grow best around 50 degrees (10 C) and spoil refrigerated foods. Others are purposely cultured in refrigerated dairy products to curdle the milk proteins. Extreme psychrophiles have been isolated from the polar ice caps and from the oceans' deepest and coldest waters. Cold turns the normally fluid parts of mesophile cell membranes into a thick gel in which enzymes stop working. Psychrophile membranes, however, contain fatty compounds that retain fluidity so their enzymes remain functioning at freezing temperatures.

The cold-loving or heat-loving bacteria do not enjoy a wider temperature range than that of the mesophiles. The entire optimal range of thermophiles is shifted toward the 120 to 230 degree F range (50 to 110 C). Likewise, the psychrophiles' range is shifted toward minus 45 to 60 degrees F (-7 to 15 C).

Temperature extremophiles have commercial value. Psychrophile enzymes are being tried in the fragrance industry to work at certain points in the manufacturing process that usually require heat. Since heat alters the characteristics of a fragrance, turning to colder-temperature biological methods of production is thought to be superior. Conversely, the thermophile *Bacillus stearothermophilus* produces enzymes that work at 130 degrees F (55 C), so is prized by detergent manufacturers for hot-water products. Thermophiles are also instrumental in ecology by composting organic wastes on farms and in nature.

Thermophiles and PCR

Thermophiles have made tremendous impact in biotechnology and are partially responsible for spawning quite a few crime-solving shows on television. Perhaps thermophiles' most famous contribution is a heat-resistant enzyme called DNA polymerase. The polymerase from *Thermus aquaticus* withstands a temperature of 160 degrees F (71 C) and therefore became a key part of a laboratory discovery by American chemist Kary Mullis in the 1980s. Mullis developed the technique known as **PCR, p**olymerase **c**hain **r**eaction, for which he received a Nobel Prize in 1993.

PCR is a way of making millions of copies of a single gene in a few hours. A small sample believed to contain a tiny amount of DNA is mixed with DNA polymerase from *Thermus aquaticus*. (The enzyme has been nicknamed *Taq* polymerase.) Nucleotides, the building blocks of DNA, are added to a tube containing polymerase and the DNA sample. The mixture is put into a thermocycler (or thermal cycler), a machine about the size of a toaster oven. The cycler heats and cools the mixture repeatedly between 162 F (72 C) and about 130 F (55 C).

During the repeated heatings and coolings, the original double-stranded piece of DNA separates into single strands each time a heating cycle occurs, called "melting" the DNA. During the heated phase, polymerase builds new strands that are complementary to the originals. The strands bind together into DNA's unique ladder shape as each cooling cycle begins. Then the heating repeats, typically for 25 to 30 cycles. At each heating, the original piece of DNA and all its new copies replicate. The total cycles therefore result in an enormous multiplication of the original DNA. This is referred to as DNA amplification. In 25 cycles, the original tiny piece of DNA amplifies to over a million-fold. This efficient system of making lots of DNA starting from a very small amount could not take place without two factors: (1) heat that causes the ladder-shaped DNA to melt or come apart, and (2) a heat-resistant enzyme that will build new DNA at raised temperature. Therefore, through use of PCR, millions of identical copies of a tiny bit of DNA are made, and this replication is done within a day.

Today PCR is used to detect HIV and other infectious agents in blood. It is also used for finding specific microbes within complicated mixtures such as soil or food. Scientists analyzing the genome

Bacteria and Cold Case Files

What do geysers in Yellowstone National Park have in common with catching criminals? The common thread is provided by thermophilic bacteria. In 1969, two Indiana University microbiologists reported finding an extreme thermophile in the 160-degree F (72 C) waters of Yellowstone's Mushroom Spring (Figure 8.1). The bacteria were assigned the name *Thermus aquaticus*. As PCR (polymerase chain reaction) technology blossomed in the 1980s, the *Taq* polymerase became its standard enzyme for efficient DNA copying at high temperatures. Years passed before PCR was applied to fields outside specialized DNA sciences. By the 1990s PCR became a hot new technology. Forensic scientists soon realized they held a valuable tool for linking suspect DNA to a crime scene.

Figure 8.1. The polymerase enzyme used in today's PCR—a rapid way to produce large amounts of DNA starting from one small piece of DNA—is from bacteria originally found in Yellowstone National Park's thermal pools and hot springs. *Copyright Ada Piro.*

On the evening of November 30, 2001, Washington State Police arrested a truck painter as he left work. After two years of legal strategizing, Gary Ridgway confessed to being the Green River serial killer, who murdered forty-eight women from 1982 to 1998 in the Seattle area. Although he was an early suspect, DNA technology of the day was not sufficient to link him to the victims. Little physical evidence was uncovered, and the cases went cold until 2001 when Detective Tom Jensen returned to the stored-evidence room. A small amount of saliva had been taken during a 1987 search of Ridgway's home. Not being a large enough sample for accurate analysis, it remained untouched for four years until Detective Jensen presumed that the current forensics might be successful where earlier attempts had failed. He sent the saliva sample to the state crime lab. There technicians used PCR to amplify tiny quantities of DNA from the killer and from swab samples taken years before from three victims. They matched. The Green River Killer is now serving forty-eight consecutive life terms without the possibility of parole.

The *Taq* polymerase was perfect for the circumstances. The enzyme requires only small amounts of DNA to begin replicating it to amounts suitable for analysis. In addition, highly degraded samples from twenty-year-old crimes are not a problem in PCR. Investigators returned in 2003 to one of the sites the killer used for disposing of bodies. Two human bones were recovered. Forensics by that time had advanced to mitochondrial PCR, a refined method for analyzing human tissues that do not have a nucleus, such as hair, bones, and teeth. Mitochondria are small energy-making packets inside mammalian cells, and they, too, contain DNA. Mitochondrial PCR therefore advances the power of biological sampling for crime solving.

At the same time that the horrific events were unfolding in Washington, a second team of intrepid microbiologists managed to excavate a sample from a Pacific Ocean "black smoker," a hydrothermal vent that is the hottest of the hot. As venting water bursts out at 750 degrees F (400 C), it meets the frigid deep-ocean water. Rather than boil—water can't boil in the intense pressures at the sea floor—the searing waters cause dissolved minerals such as iron to chemically bond with sulfide emerging from the vent. The resulting black iron monosulfide billows forth, giving these thermal vents their nickname (Figure 8.2).

Thermococcus litoralis was discovered in black smokers. Because it is an extreme thermophile, scientists were excited about using its polymerase in PCR. The *T. litoralis* version turned out to be more efective than Taq polymerase. While Taq polymerase will copy any piece of DNA, even a piece containing a small error, *T. litoralis* polymerase has proofreading ability. It also works at higher temperatures than Taq. *T. litoralis* polymerase may mark the next generation in PCR capabilities, not only for medical science but also for chipping away at law enforcement's daunting backlog of cold case files.

Figure 8.2. Bacteria living in the caustic environment of black smokers on the sea floor may be the most extreme examples of thermophiles. Their enzymes are now being applied to PCR methods. *Copyright William J. Brennan/ SEPM.*

(the entire collection of a species' genes) of animals and plants use PCR to amplify specific genes. The extraordinary speed with which the human genome project characterized human DNA could not have happened without PCR.

Solving crimes through physical evidence also depends on the PCR technique. DNA is in whole blood, urine, saliva, semen, hair, tissue, bone marrow, and tooth pulp. (It is not present in red blood cells, which do not have a nucleus.) No matter how degraded or aged the sample, DNA can be amplified to amounts large enough to be accurately compared with a specimen taken directly from a suspect. Forensic scientists identify the correct suspect when a gene from the suspect exactly matches one from the DNA recovered at a crime scene. Additional matches between genes simply increase the certainty.

Acidophiles and Alkaliphiles

Acidophiles are microbes that live in very acid conditions, sometimes at a pH of less than two. Acidophiles are used in making foods such as vinegars and sauerkraut, and they help preserve foods by making the conditions too acidic for most other bacteria.

Drainage from ore mining operations is acidic and damaging to the environment as it leaches into soils or runs into streams. *Thiobacillus ferrooxidans* bacteria contribute to this acid production. The copper industry uses the microbe's acids to leach metal from low-grade ore that remains after coal is mined. Before this type of biomining was perfected, the mining industry lost millions of dollars worth of metal that was otherwise too expensive to extract. Biomining has since become a multibillion-dollar business.

Like acidophiles, **alkaliphiles** are specialists that keep their cell interior neutral even when living in an extremely harsh alkaline environment. Such conditions are found in the world's salt lakes. Alkaliphiles are used in leather treatment and in remediation of areas contaminated by oil spills.

High Salt Environments

The alkaliphiles are also **halophiles,** microbes that live in high salt solutions. Halophiles must find ways to tolerate both the salty surroundings and the pressure difference between the outside and the inside of their cell. Most microbes can survive for a time in sea water, which has salinity about six times that of their optimal salt level. As salt enters cell contents, the cell's internal pressure rises. To survive in very high salt concentrations, halophiles employ tactics to keep the salt outside the cell and maintain an interior that is 80 to 90 percent water. Extreme halophiles *require* high salt conditions. The bacteria living in the Dead Sea, for instance, need salt levels at 30 percent.

Halophiles are not confined to the seven seas. They live in salt flats and even on the salt licks used on farms for cattle and horses. Some have been isolated from brine salts and are used in the brine-curing of animal hides.

Halobacterium halobium, native to salt lakes, produces a purple protein called bacteriorhodopsin that is similar to the pigment in the eye for vision in low light. The light-sensitive properties of bacteriorhodopsin are currently being studied for use in biochips. These are computer chips that process data, but they use biological compounds rather than silicon. *Halobacterium*'s purple protein carries information at the speed of light, outdistancing silicon in processing speed. The advancing field of artificial intelligence may someday soon employ bacteriorhodopsin biochips.

Other Extreme Conditions

Technology has made use of select thermophiles, psychrophiles, and acidophiles. The advantages of other extreme-loving microbes have not yet been fully realized, but it seems likely there will come a day when their specialties will benefit society.

Barophiles live in high-pressure habitats as found in the deep sea. The barophiles on the ocean floor tolerate pressures 1,000 times that of the atmosphere. Other microbes of the "deep" have recently been discovered in sunless mines two miles below the earth's surface.

Astrobiologists study psychrophiles and barophiles to answer questions on planetary life other than earth's. The cold temperatures and high pressures favored by these microbes resemble those found in our galaxy. These specialized microbes therefore serve as a model for the biology of the solar system and beyond.

Some bacteria manage to exist in an environment that gives them next to nothing in nutrition. Ultrapure water used in the semiconductor industry for rinsing circuits is distilled and repeatedly filtered to eliminate solids and particles, salts and minerals, organic compounds, and silica. But *Caulobacter* and *Pseudomonas fluorescens* are two bacteria that have the potential to disrupt the manufacture of electronics. These bacteria are able to extract enough nutrients from the purified water and carbon dioxide from the atmosphere to meet all their needs for growth.

Additional specialists of the microbial world are the phototrophs, microbes that need only light to supply their energy, and autotrophs, which use only carbon dioxide for metabolism. Autotrophs are also called lithotrophs, or rock-eating bacteria, because of their seeming indifference to rich nutrient sources.

Xerophiles are microbes that live in places with almost no moisture. Fungi are more apt than bacteria to live in the xerophilic conditions of deserts. Some xerophiles spoil dry stored grains, seeds, and nuts. In food products, high sugar content or salt is often used for preservation by reducing the water available for microbial growth. Unfortunately, xerophiles live well in sugary or salty places so cause trouble in foods that are inhospitable to other microbes.

Certain microbes living in the most extreme of the extreme environments on earth can be quite scarce. Their habitat is called the **rare biosphere**, a place so distinct and unusual that life there often does not resemble anything else we know. In time, microbiologists may find rare microbial species that conduct valuable biological reactions. For now, simply finding and recovering these microbes is the greatest challenge.

Bioremediation

Superbugs

Biological methods for cleaning up environmental pollution depend on extremophiles. Their ability to live on toxic dumps of metals and organic solvents makes them valuable aids to detoxification of pollutants. To date, environmental chemists know of over 1,000 chemicals that are destroyed by some type of microbe. Considering that 20 million Americans live within four miles of EPA-designated Superfund sites, there is no surprise that biological recovery of ecosystems is a growing industry. Microbes are now used or are being investigated for cleaning surface and deep soils, sediments, groundwaters, surface waters, oceans, estuaries, and wetlands. Detection instruments are being refined each day for measuring minute amounts of pollutants in solids and liquids. As analytical sensitivity improves, more contamination is revealed and in an alarming number of places.

Many microbes that are naturally found in soil and water slowly degrade or neutralize (turn into nontoxic forms) the chemicals dumped into their surroundings. Natural degradation takes more time than our environment can afford. Toxic site cleanup therefore will need to depend on bioengineered microbes to speed up the process.

To "invent" a bioengineered microbe, a microbiologist first finds a species that is growing in the polluted site (Microbe #1). In order to survive in toxic soil or water, some extremophiles have enzymes that act upon hazardous solvents or metals. The microbiologist studies the enzyme as the microbe produces it in a test tube or petri dish. He then locates the enzyme's gene in Microbe #1's DNA. The chosen gene(s) is transferred to a second microbe (Microbe #2) by a process called **gene transfer.** In this way, Microbe #2 is turned into a bioengineered superbug with an appetite for pollutants.

There are various gene transfer options. The first is **conjugation** in which two cells in contact with each other exchange DNA. A second method called **transduction** moves genes from one type of bacteria to another by using viruses that infect only bacteria (bacteriophages). The virus is made to carry the desired gene and then infects the target bacteria. In this way, the gene is inserted into a new superbug's DNA. Microbiologists also use **transformation** to get a gene from one bacterial type to the next. In transformation, "naked" DNA containing the important gene is put into a liquid and then bacteria are added. The bacteria drink up the DNA and incorporate the gene into their own chromosome. Finally, genes may be transferred by using **electroporation.** Cells and DNA are both put into a liquid, and then an electric current is applied. The current causes tiny pores to open on the bacteria, and DNA passes through them and into the cell. Plasmids may also be used for carrying desired genes from one microbe to another. Most of these methods work on some bacteria and not others, and all involve a few more manipulations than are mentioned here.

Once a gene-transfer method is chosen, a microbiologist selects recipient bacteria based on their hardy growth characteristics. Spore-forming *Bacillus* is a popular choice, as are *Pseudomonas* and *Alcaligenes,* which grow well in many environments. *Penicillium* and *Fusarium* fungi are helpful, too, in some instances. Superbugs combine natural robustness with the special gene now part of their DNA. If a superbug is made correctly in a lab, it does not merely survive on a chemical pollutant; it craves it.

Before long, bioengineered superbugs will be a standard part of toxic site cleanup. Superbugs are being developed to target some of the following organic chemicals: benzoate, toluene, naphthalene,

It's Dirty Work, But . . .

In 2006, CareerBuilder.com identified the "Ten Dirtiest Jobs in Science." Microbiologists have the dubious distinction of being responsible for five out of the ten jobs. The list includes **manure inspector,** orangutan urine collector (collects output for reproduction studies), **hot-zone supervisor, extremophile excavator, dysentery stool analyst,** semen analyst (counts sperm cells and preserves for *in vitro* fertilization), volcanologist (monitors active volcanoes), carcass cleaner (prepares slaughtered livestock for the meat production line) **rumen fistula specialist,** and corpse-flower grower (grows and tends to the foul-smelling plants).

MICROBIOLOGY JOB	WHAT THEY DO
Manure Inspector	Paw through farm manure to ensure it is free from contaminants before use on crops.
Hot-Zone Supervisor	Maintain specialized labs that study the most deadly pathogens, i.e., anthrax.
Extremophile Excavator	Sift through Superfund sites, steaming springs and vents, and below freezing locations to find specialized microbes.
Dysentery Stool Analyst	Study pathogens in stool from diseased persons to enlarge our medical knowledge.
Rumen Fistula Specialist	Reaches inside a cow's stomach through a surgically implanted opening (fistula) to study rumen microbes and digestion.

octane, ether, and creosote. This is but a short list. Some species are known to degrade over 200 organic chemicals. Bioremediation has already been successful in cleaning gasoline and fuel oil spills and pollution from chlorinated organic compounds.

Biochips

In at least one instance, the world of computers is joining with biotechnology to detect environmental toxins. Luciferase is an enzyme that causes some biological things to emit light. Summertime in parts of North America would not be complete without small green flashes in the darkness, the telltale sign of fireflies' luciferase. On a nighttime ocean cruise, it produces the eerie wake of phosphorescence from trillions of luminescent plankton floating in the waves.

In technology circles, microbial luciferase may become a type of on-off switch for signaling the presence of pollutants. The idea is to combine the conducting features of a processing chip with the biological activity of the enzyme. The result is a **biochip.** Certain compounds that react with a contaminant in soil or water give off a small burst of energy. If these compounds are attached to a chip or probe along with luciferase, the luciferase would emit a flash of light each time the chip detects a toxin molecule. The biochip may even be designed so that the amount of light emitted is proportional to the amount of pollution.

Bioremediation Promise and Questions

Metals' contamination of the environment can come from natural activities that disrupt geological structures. Earthquakes and volcanoes are examples. The electrical, paint, alloy, nuclear, and mining industries also produce metal by-products. Superbugs that attack metal pollutants are produced in the same way as they are for organic solvent degradation. First, microbes living in close association with the metal are recovered and brought to a laboratory. Scientists study the methods each species uses to protect it from the metal's toxicity. The hardiest microbes are engineered to express super-detoxifying activity. A mixture of the most efficient superbugs is then inoculated into the contaminated site. The microbes remove the toxic threat in one of two ways: (1) by immobilizing the metal so it cannot leach into ground or waters, or (2) by binding the metal in a complex created by the microbe. Hazardous-waste technicians then physically remove the bound metal. Trials are under way to use superbugs to target today's prevalent metal contaminants: selenium, arsenic, cadmium, mercury, manganese, zinc, nickel, and lead.

In addition to superbugs, biofilms are cleaning waters that contain hazardous metals. Biofilm microbes pull out the metals as they stream past. Remedial biofilms are like naturally occurring biofilms—they are made of a diverse mixture of microbes—but they are manipulated to contain microbes with a hunger for pollution. For metal-contaminated waters, certain bacteria seem to work best: *Bacillus, Citrobacter, Arthrobacter,* and *Streptomyces.* The yeasts *Candida* and *Saccharomyces* also help in remedial biofilms.

Bioremediation holds the potential for arresting some of the damage inflicted on ecosystems every day. There are critics who fear the accidental release of bioengineered microbes into the surrounding environment, where the superbugs might interfere with natural ecosystems. Because of strong backlash against genetic engineering, the use of superbugs has been slow to come on the scene. Bioengineered microbes are used sparingly on toxic sites around the world. The 1989 Exxon *Valdez* oil spill cleanup used a low-risk version of bioremediation. Bioengineered microbes were not called upon in the initial months after the spill. Instead, nutrients were added to the contaminated shores, the nutrients accelerating the growth of native microbes already growing on the oil. This approach is called **intrinsic bioremediation** because it uses biological factors already present in the contaminated stream, shoreline, or soil.

Many people have the same concerns over bioengineered microbes for toxic cleanup as they do for bioengineered microbes for agricultural use. What is the real risk of a bioengineered species escaping its intended confines? If it does escape, what damage will it cause by entering our existing ecosystems? Risk-analysis scientists blend mathematics, probabilities, and the statistics of actual events to assess the health risks of accidents.

Although the thought of a genetic monster released into society makes for exciting movies, a number of events would have to take place in sequence to initiate a biological disaster. The events leading to biological disaster would have to proceed in order: (1) the superbug would persist in an environment outside its intended target area; (2) it would then find a way to maintain its numbers within a population of native microbes that are already adapted to the environment; (3) it would find avenues for further dispersal; (4) its artificial genetic makeup would harm hosts such as fish, animals, or humans; and (5) it would establish a niche that allows it to multiply and spread in the host population. The likelihood of all of these things happening in a specific sequence is small.

There, too, is the comfort of the **dilution effect.** This natural safety mechanism works equally well against superbugs and water-contaminating bioterrorists. The bioengineered superbug released into the world must overcome the natural processes that have developed over millennia. Although the accidental release of superbugs is possible, their chances for success are small because other microbes and environmental conditions would overwhelm their actions.

Still, predicting the vagaries of biological systems is always a dicey affair. The threat most likely does not come from a bioengineered microbe properly applied and monitored. The biggest risk of a bioengineered catastrophe comes from human error, of which there seems to be no shortage.

Higher High Tech
Gene Therapy
Vaccines were a medical breakthrough when Edward Jenner first tested them in 1798. The principles of vaccine production have not changed much since then. In the past decade, however, a second major advance in health care has emerged from the biotechnology

world. It is **gene therapy,** and it takes advantage of viruses' ability to invade host cells.

In gene therapy, viruses act as vehicles for delivering genes to specific tissues in the body. A normal gene replaces a damaged or abnormal one in the chromosome, or a gene is inserted to replace one that is missing. Viruses are biology's most efficient cellular invaders, and once inside the cell they take over normal replication of host DNA. Gene therapy's potential is in curing genetic diseases at the point of their origin, DNA.

FDA has not yet approved gene therapy for sale. The first gene therapy trial was conducted in 1990, and, as with any budding science, there are challenges to overcome. Nevertheless, this therapy has shown promise in treating patients with certain immunodeficiencies, muscular dystrophy, and cystic fibrosis.

At present, researchers are working to improve aspects of gene therapy before it will be fully effective. They must first find ways to induce host DNA to permanently retain the newly inserted gene. In addition, the body doesn't differentiate between "good" therapeutic viruses and "bad" viruses. Since any virus in the bloodstream is a foreigner, the body reacts by setting up inflammatory and immune responses, which must be suppressed when injecting the therapeutic virus. There is evidence of and there are fears about therapeutic viruses reverting to their infectious form. Finally, many illnesses are controlled by more than one gene. Diabetes, heart disease, high blood pressure, Alzheimer's disease, and arthritis are multiple-gene disorders. Gene therapy may not be the optimal choice for treating some of these diseases. Conversely, if a condition is controlled by just a few genes, then gene therapy shows promise. Current research in gene therapy is focusing on sickle-cell anemia, hemophilia, insulin replacement for type 1 diabetes, genetic hypercholesterolemia, and specific tumors.

Nanobiology

Nanotechnology is the science of building devices so small they can manipulate molecules inside a cell. The scale of nanotechnology ranges, but it is generally less than 1,000 nanometers. Nanobiology combines electrical or nonelectrical instruments, termed nanodevices, with a biological component. The nanodevice may be a nanowire, a microcircuit, or an electrode. Potential benefits from nanobiology include drug delivery, disease diagnosis, or tracking the progression of a disease inside cells.

Nanobiology techniques are being used for building artificial layered membranes based on the structure of bacterial membranes. These double-layered membranes form a bag called a vesosome, which is more likely to avoid destruction by the body as it carries drugs to diseased organs and tissues.

A virus known as M13 holds promise for nanobiology because it binds with metals and, in large groups, forms organized sheets over inert surfaces. Assemblies of M13 may be utilized to conduct electrical current and perhaps signal the presence of a toxin inside a human cell. If a single toxin molecule is known to induce mutations in DNA, nanobiology may become the first step in predicting the onset of cancer.

Microbes in Society

The world's population will pass 6.5 billion by the time you finish reading this book. As the population grows it migrates toward urban centers. Urbanization puts discrete stresses on both the human body and its surrounding ecosystem. We have belatedly realized the alterations humans have made to the climate, the oceans, the atmosphere, and natural resources.

Microbes are not detached from these changes. Infectious agents will have increased opportunity to find susceptible hosts in densely populated communities. A significant percentage of people in these communities may be people with high-risk health conditions. Polluted environments will further increase our exposure to toxic chemicals in food, water, and air, and infectious microbes will take advantage of the stresses on human and animal health. The rate of new disease emergence may increase as our planet fills with toxic waste.

However, mammalian life on earth would not exist for long without microbes. Their roles in the biosphere are almost endless: waste decomposition, nutrient recycling, vitamin production, food digestion, food production and preservation, and participation in the manufacture of industrial and consumer products (Figure 8.3).

Most of earth's microbes are an untapped resource for enzymes, proteins, antibiotics, and chemotherapeutic drugs. Their potential to cure illness and detoxify our living space far exceeds their threat to us. Extremophiles are virtual unknowns in the microbial world and may eventually provide answers to undiscovered parts of earth's biosphere and those beyond our planet. The limiting factor in microbiology may be society's resistance to new technology. Each leap in scientific discovery throughout history has struggled through a period of negative backlash. Genetic engineering, nanotechnology, and gene therapy bring inevitable arguments directed against the risks of manipulating natural systems. Some arguments may have validity. The dire condition of our landscape is, after all, the consequence of technology.

It would not be a mistake to study microbes to learn the ways living cells can adapt to any environment. A mistake humans make is to believe we are separate from all other

Figure 8.3. Attention to detail and strong eyes (for hours at the microscope) are required for microbiology. Many microbiologists must also be able to seek samples in uncomfortable and foul-smelling locations. *Copyright 2006 ATS Labs and Voyageur I.T.*

livings things on earth. A worse mistake would be to think that we are meant to dominate all other creatures. Humans are an equal member of our planet's biota, sharing space with the plants and animals and microbes around us. If the case for one creature's superiority over another were to be made, it would not be humanity that wins. For adaptability, the capacity to "outwit" agents designed to destroy them, and readiness to thrive despite physical and chemical obstacles, the microbes will prevail. They existed on earth long before humans, and there is no doubt they will populate the planet once humans are gone. If you elected to choose a superior organism in biology, you could do no better than to pick the single-celled dynamos, the microbes.

THE FIVE-SECOND FINALE

The Five-Second Rule is based on core principles of microbiology. Take time to review those principles you have learned here before you reach down to pick up the cookie you dropped.

Microbes are everywhere, so assume that the cookie will have picked up a few or a few dozen from the floor, especially since you now know that **microbes can be transmitted indirectly from person to person** on inanimate objects. Five seconds is plenty of time for this transmission from floor to cookie. Examining the cookie and reasoning, "It looks clean," doesn't help because **microbes are invisible**. Fortunately, **most microbes are not pathogens**. If there are one or two pathogens on your cookie, **your immune defenses will defeat small infections**. Anyway, what are the chances that the cookie picked up an **infectious dose** of a disease-causing microbe?

You may now weigh your choices based on science, rather than making a decision based solely on the amount of chocolate chips in the recipe. But maybe there are other factors working here. Perhaps the Five-Second Rule isn't about microbiology, after all. The University of Illinois once made a detailed study of the rule. One finding may not surprise you. The scientists learned that people were much more likely to pick up and eat cookies and candy off the floor than cauliflower or broccoli!

TWENTY-FIVE FREQUENTLY ASKED QUESTIONS

1. *Can a bar of soap hold germs?* Yes, used bar soaps can have bacteria on their surfaces. It shouldn't be surprising; bacteria live almost everywhere on earth, and soap ingredients plus a watery environment provide good growth conditions. Microbiologists have simulated the amount of soap transferred to the hands during a typical hand washing, and determined that the bar soap may carry from a few dozen to 10,000 bacteria. Most of them are the common skin bacteria *Staphylococcus aureus*. Liquid soaps dispensed from bottles contain less bacteria than bar soaps, though the pump handle and nozzle are likely to be contaminated. The amount of bacteria on a bar of soap is far less than found on your hands. Thorough hand washing removes most of the bacteria on hands and on the soap. The quality of your hand washing is more important than any potential bacteria on the soap. Hygiene experts have suggested using the Double Happy Birthday approach: wash your hands through two choruses of "Happy Birthday to You." (Resources: *Applied and Environmental Microbiology* 48:338, 1984; *Epidemiology and Infection* 101:135, 1988; *Infection Control* 8:371, 1987)

2. *Isn't getting a vaccine more dangerous than not getting one?* No. Hundreds of millions of lives have been saved worldwide due to

effective vaccination programs. Deaths from the complications of a vaccine occur, but are rare. Therefore, vaccines are an overall health benefit. The health concern regarding vaccines centers on the possibility of the vaccine causing (a) infection or (b) an allergic reaction. Attenuated (weakened) vaccines use live viruses that have been treated so they cannot cause infection, but some are known to cause side effects. The attenuated MMR—measles, mumps, rubella—vaccine is an example. Up to 15 percent of persons receiving MMR vaccine may develop low-grade fever and/or rash about a week after the injection. The CDC suggests that egg protein–based vaccines (MMR, influenza, yellow fever) pose too high a risk of an anaphylactic reaction in people with allergies to egg proteins, and such vaccines are to be avoided by those people. (Resources: National Immunization Program of the Centers for Disease Control and Prevention; National Vaccine Information Center; *Morbidity and Mortality Weekly Report,* December 1, 2006)

3. *Is my sponge really going to make me sick?* It's possible. The easiest way to get sick from using your dirty sponge is by using it to soak up liquids or blood from raw meats, then immediately using the same sponge to wipe a surface that contacts raw vegetables or fruits, or is used for slicing bread. Always give those kinds of shared surfaces a thorough scrub with soap and warm water in between uses. Disposable towels are a better choice than sponges for wiping up contamination from raw meat or fish. If sponges are used for this task, replace them or decontaminate them. (Resources: Washington State University Extension Service; Community Practitioners' and Health Visitors' Association)

4. *Who are cleaner, men or women?* A loaded question. Studies comparing restrooms and hand-washing habits show that women and men have different hygiene habits—no surprise—and in least one

area, women are microbiologically dirtier than men. Microbes are more numerous in women's restrooms, and more surfaces have contamination than the surfaces in men's restrooms. Women do better in hand-washing habits. A Harris Interactive survey in 2005 watched the habits of men and women in several public restrooms across the United States. Ninety percent of the women observed washed their hands, but only 75 percent of men washed. Other studies have shown the percentage of hand washers to be even lower. Neither men nor women do well in honesty. In a sister survey done by phone, 97 percent of women and 96 percent of men say they always or usually wash their hands after using a public restroom. (Resources: Charles Gerba, PhD, University of Arizona; The American Society for Microbiology; The Soap and Detergent Association; Harris Interactive Inc.; *New York Times,* February 23, 1999)

5. *Do antimicrobial soaps really work?* Yes, depending on how they are used for washing. There is scientific evidence that antimicrobial soaps may work better than regular soaps in reducing the numbers of bacteria on hands. The drawbacks to these studies include the subjects, who tend to be people from the health professions who are more aware of proper hand-washing technique than people not trained in hygiene. Another aspect overlooked in many hand-washing studies is the manner in which average Americans go about the washing process every day. Most people do not wash their hands long enough or at the proper water temperature for soap, any soap, to have its best effect. In those cases, the antimicrobial ingredient in soap may not kill or reduce many microbes. In conclusion, proper hand washing with *any* type of soap is recommended for reducing the spread of germs. (Resources: *Journal of Community Health* 28:139, 2003; *Infection Control* 8:371, 1987)

6. *Can mosquitoes carry AIDS?* No. The AIDS virus, HIV, is transmitted through sexual contact or through parenteral routes (direct blood transfer) such as dirty needles. The reasons are the following: (a) HIV does not replicate inside mosquitoes and transmission by way of vectors must involve virus multiplication inside the insect vector. (b) HIV does not survive for long inside mosquitoes because insects do not have CD4 lymphocytes (CD4 antigen on lymphocytes is required for virus attachment to cells), which is the cell HIV targets for infection. When mosquitoes digest blood, the ingested viruses are destroyed. (c) Mosquitoes inject saliva when biting a host, not blood. (d) There have been exhaustive studies in parts of the world with very high AIDS rates and a large insect vector population. Vector transmission of HIV in doses high enough to cause AIDS has never been proven. (Resources: The Centers for Disease Control and Prevention; *Journal of the Louisiana State Medical Society.* 151:429, 1999; Rutgers University Cooperative Research and Extension; Los Angeles County West Vector and Vector Borne Disease Control District)

7. *Some sanitizer sprays say they kill odor-causing bacteria in the air. Do bacteria in the air really cause bad odors? Or does the spray just cover up odors with a fragrance?* Bacteria in the air do not cause odors. Airborne bacteria are usually in moisture droplets or on items that travel on the breeze: dirt, dust, pollen, leaves, hair, and so on. During that brief flight time, bacteria digest few nutrients and do not give off bad odors. Once they settle on a surface, however, they may grow and eventually cause odor. Most air sanitizers sold today are not tested on bacteria; the sanitizer claim is based on the product's ingredients, which often includes a fragrance. (Resource: The Environmental Protection Agency; Reckitt-Benckiser Basic Microbiological Control Manual)

8. *Is washing hands with hand sanitizers just as good as soap and water?* Both are helpful, either one is essential. Eighty percent of all infectious diseases are transmitted by human contact, and much of that is via the hands. Soap and water removes dirt, hair, dead skin cells, and many microbes, and should be done before food preparation and eating, after diapering a child, after handling pets, after being outside, and after using a restroom. Hand sanitizers use alcohol, which is an effective antimicrobial substance that is irritating to the skin for some people. Hand sanitizers are recommended when soap and water is not immediately available in such instances as air and other types of travel, sporting events, camping or outdoor events, and so on. At least one study has shown that alcohol gel sanitizers are more effective than soap and water for removing germs from the hands. (Resources: *American Journal of Infection Control* 27:332, 1999; Charles Gerba, PhD, University of Arizona; Philip Tierno, PhD, New York University Medical Center)

9. *Is a dog's mouth really cleaner than a human's?* A dog's mouth is different from a human's, but not cleaner. Dogs have a large amount of oral bacteria. Their bacteria differ from that of humans, possibly because of diets low in carbohydrates and different salivation patterns. Dogs very rarely get cavities, but they do suffer from gingivitis, which can turn into periodontal disease and tooth loss if left untreated. Plaque forms readily on canine teeth. Dogs inoculate themselves with fecal bacteria by licking the urinary and anal areas, skin and coat, and paws, which can pick up an untold amount of microbes. Some dogs also practice the charming habit of dung eating, medically known as coprophagia. Perhaps people think canine mouths are clean because humans don't catch colds or flu from dog kisses as they could from human admirers. (Resources: *The Merck Veterinary Manual;* Douglas Island Veterinary Service LLC; Hilltop Animal Hospital)

10. *Is it "Feed a cold and starve a fever"? Or, "Feed a fever and starve a cold?" Either way, how does this help kill viruses?* Not only is there debate on the correct wording of the proverb and its original meaning but also questions persist on the medical merits of doing either one! The correct modern phrase is "Feed a cold and starve a fever," and originates from a similar English saying used in the 1500s that likely did not carry the same meaning as it does today. Most medical experts have for years agreed that adequate nutrients, fluids, and rest are the best things for colds and flu, and that fasting when you're sick is a bad idea, so the proverb was dismissed with other medical myths. Then in 2002, a team of Dutch medical researchers found evidence that eating a balanced meal increases the body's production of a compound (gamma interferon) that destroys viruses. Fasting stimulates a compound (interleukin-4) that fights the bacterial infections that cause fever. In other words, the old saying may provide sound medical advice after all. The Dutch study was very small, using only six male volunteers, and has not been repeated in larger studies. Therefore, until further notice, when you have a cold or the flu, get plenty of bed rest, drink plenty of fluids, and enjoy a warm bowl of chicken soup. (Resources: Allina Health System; Indiana University School of Medicine; Cardiff University; *Medical Hypotheses* 64:1080, 2005; *Clinical and Diagnostic Laboratory Immunology* 9:182, 2002)

11. *Are we doing more harm than good by constantly using disinfectants and sanitizers?* This argument has not been resolved. Evidence shows that a regular schedule of disinfecting the home lowers the pathogens found there. The benefit is short lived, however, due to reuse of the cleaned surfaces. Opposing scientists argue that chemical disinfection leads to resistant microbes and hinders the chance for normal immune system development. Both schools of thought are accumulating libraries of scientific articles that support their

arguments. Because emotions run high on this question, it is becoming increasingly difficult to objectively sort through the data. (Resources: *Journal of Applied Microbiology* 85:819, 1998; *Journal of Applied Microbiology* 83:737, 1997; Reckitt Benckiser plc, The Clorox Company; Alliance for the Prudent Use of Antibiotics)

12. *Can I catch anything from a toilet seat?* Not unless you really try. Toilet seats are relatively clean compared with other objects in the bathroom and even the kitchen. A toilet seat is like any other inanimate surface that is shared by many people, most of them healthy. You catch dangerous microbes from any surface that has obvious filth on it, but you can also pick up pathogens from a seemingly clean surface. To reduce the risks of carrying microbes from the bathroom around with you all day, wash your hands thoroughly with soap and warm water for at least twenty seconds after using the bathroom, *every* time you use a bathroom. (Charles Gerba, PhD, University of Arizona; *Where the Germs Are* by Nicholas Bakalar, 2003)

13. *What is the best way to disinfect a dog kennel?* Remove bowls, beds, toys, and the dog! Remove all visible dirt, and then scrub all surfaces with water and a robust brush. Rinse thoroughly with hot water. With a sprayer or mop, apply a bleach solution or other disinfectant solution—never combine them—according to the product's directions. For concrete-floor runs, use only products that state they are effective for these types of surfaces and follow the directions. For hard, nonporous surfaces such as hard plastic or metal cages, 3/4 cup regular household bleach to a gallon of water is recommended (1 part to 22 parts water). Cover all surfaces and let the bleach solution remain wet on the surface for at least ten minutes. (For nonbleach products intended for concrete dog runs, follow the application instructions and contact time stated on the label.) Rinse thoroughly with water by hosing, and dry the kennel as

completely as possible with a squeegee and ventilation. Because disinfectant fumes are irritating to animals, rinsing well is a top priority, even for products that state rinsing is not necessary. Bleach is corrosive to metal surfaces, and many kennel owners alternate its use with other nonbleach disinfectants on an every-other-day or weekly schedule. (Resource: Humane Society of the United States)

14. *What should I do about anthrax?* The Department of Homeland Security Web site provides links to several advisories and discussions on anthrax bacteria and associated risks. The Centers for Disease Control and Prevention Web site provides question-and-answer pages on anthrax, as well as advisories for contacting law enforcement if you suspect you have been exposed to anthrax. All agencies advise individuals to report any suspicious package or envelope that appears to have on it a powder of any color.

15. *How long does a cold virus live on a doorknob?* There are few absolutes in biology. Some cold viruses remain active on a surface like a doorknob for minutes. But there is evidence that cold viruses can persist for up to three days on hard, inanimate surfaces. After touching any shared object such as kitchen counters, faucets, and refrigerator handles, do not touch your hands and fingers to any part of your face before washing. (Resources: Syed Sattar, PhD, University of Ottawa; The Centers for Disease Control and Prevention)

16. *Will there come a day when all bacteria are resistant to all known antibiotics?* This is unknown, but it probably won't happen. There are hundreds of thousands of microbes in the world yet to be discovered. Some of them may in the future cause new diseases that humans have never seen before. These new species will probably be susceptible to antibiotics, though we know resistance in microbes can develop quickly. There, too, is a vast resource of undiscovered

plants and microbes that might produce new and effective antibiotics. Though the known pathogens prevalent in hospitals and in the general population are becoming increasingly resistant to current drugs, there is also the belief that technology and new discoveries will keep us just ahead of the threat.

17. *Is a cold transmitted before the sick person shows symptoms?* When cold viruses infect you, they begin replicating in the mucous linings of the nasal passages before symptoms appear, which happens about a day later. As the nasal virus load increases, you can begin spreading the virus by way of your hands or saliva. It may be more common, however, that colds are spread when symptoms are in their full fury. Wiping a runny nose, contaminating your hands, and then handshaking or touching shared surfaces is known to transmit cold viruses. Sneezing, runny nose, and coughing are excellent warnings that a person is a cold factory. (Resource: The Common Cold Centre of Cardiff University)

18. *Is it safer to enroll my child in day care or to keep him home and away from germs?* This is another area in microbiology that is debated. Day care is safe when parents and the day care staff are aware of the need for good hygiene. An extra degree of attention to cleanliness and personal hygiene is necessary when young children gather and share toys, crawl on the floor, and put fingers and toys into their mouths. When elementary school children practice good hand hygiene, they miss about two and a half days of school on average in one year due to illness. Children who do not use good hygiene miss a little over three days due to illness. Adults must have good hygiene and then help toddlers stay clean. Keeping sick children home is essential. Finally, child care professionals, the center's cooks and cleaning staff, *and* parents must be educated in good hygiene. It is safe to use day care when these aspects are addressed.

There are germs at home, too. (Resources: *American Journal of Infection Control* 28:340, 2000; *Epidemiology and Infection* 115:527, 1995; Pediatrics 94:991, 1994; *American Journal of Epidemiology* 120:750, 1984; The Centers for Disease Control and Prevention)

19. *How long can I go without taking a shower?* It depends. How many friends do you want? Skipping showers affects your aesthetics more than it affects your health. This is true only if you do not receive a cut, abrasion, rash, or other break in your protective skin barrier against germs, which can lead to infection.

20. *Are hotel rooms full of germs?* Yes. Hotel rooms have microbes on almost every surface and often in very high concentrations. Though hotel rooms are cleaned, the cleaning staff has time constraints preventing a thorough cleaning and disinfecting. Some items in hotel rooms are handled by many people, and these items are never disinfected: bedspreads, telephones, refrigerator/microwave handles, toilet handles, TV remote controls, and video game controls. Many travel outfitters sell portable black lights that you can use to find the germs in your hotel room. These lights cause natural phosphor compounds to glow. Be aware, however, that things in addition to microbes contain phosphors: feces, semen, sweat, saliva. You may never stay in a hotel again after you use a black light. (Resources: Charles Gerba, PhD, University of Arizona; MSNBC.com, September 29, 2006; ABC News, January 15, 2006)

21. *Can you catch anything from a telephone?* Microbes cover inanimate surfaces such as phones. The mouthpiece collects germs from hands and from the speaker's mouth as they talk. Touching your hands to your face immediately after using a telephone or cell phone will transmit germs from the phone to you. Colds and flu are transmitted on nonliving objects and surfaces. After an hour on a hard,

dry and inanimate surface, up to 40 percent of cold viruses remain infectious. (Resources: Syed Sattar, PhD, University of Ottawa; *Washington Post,* January 11, 2006; WebMD Medical News, June 23, 2004)

22. *How safe is sushi?* Thousands of diners eat sushi every day and live to see the next sunrise. By contrast, cruise ship meals, salad bars, and fast-food hamburgers make up a gastronomic minefield. Even the symbol of a healthy diet, fresh vegetables, seems hazardous now that *E. coli* O157 and 0126 have come on the scene. Sushi restaurants are like other restaurants. Check them for cleanliness as best you can before you order, and keep an eye on the hygiene used by servers and the cooks. Eating raw meats and seafood brings an extra degree of risk because they are more likely to contain microbes and parasites. Patronize only those sushi restaurants that have exemplary reputations and pass your county's restaurant inspection report. The FDA recommends that people with high-risk health conditions, such as those with weakened immune systems or with liver disorders, avoid sushi and sashimi. (Resources: FDA Center for Food Safety and Applied Nutrition; University of Texas–Houston Medical Center)

23. *I get sick every time I fly. What can I do?* You may be catching an infection in places other than the airplane. Long-distance travel, waiting in crowded terminals, vacations, family get-togethers, and business meetings also give germs opportunities to spread. During flight your chances of catching a "bug" on the airplane increase as the flight's hours in the air increase. In flight, avoid as many shared items as possible; magazines, tray tables, pillows, blankets, and earphones can carry microbes. Some people use hand sanitizers or wear masks to keep germs away. If possible, take flights that are less crowded. In the terminal, stay away from crowds. Always wash your

hands with soap and warm water before the meal (the hand sanitizer is handy for this, too) and after each use of the restroom, including the restroom in the terminal. (Resources: World Health Organization; *Wall Street Journal,* January 6, 2006; *The Secret Life of Germs* by Philip Tierno, 2001)

24. *Do microbes grow faster in the summer when it's hot than they do in the winter?* Yes they do, but only in a generalized way. The microbes that grow best in the temperature range that suits humans slow down as it gets colder. They stop growing altogether when they are frozen. Outside on a hot summer day, microbes chew on compost piles, turn birdbath water green, and even decompose dead bodies faster than in the cold of winter. The microbes on and in your body enjoy generally stable and warm temperatures. They grow as well in the winter as they do in the summer.

25. *Is it okay for my dog to drink out of the toilet bowl?* It's as okay for your dog to imbibe as it is for you. Cleaning product residues and disinfectants released from cleaning tablets placed in the tank can lead to problems from mild stomach irritation, nausea, and vomiting, to severe gastrointestinal illness. The bacteria in toilet water can make dogs sick, too. *E. coli* sickens dogs just as it does humans. The prevention for toilet-water drinking is simple. Close the lid. (Resources: American Society for the Prevention of Cruelty to Animals; American Animal Hospital Association; American Veterinary Medical Association)

GLOSSARY

Acidophile—A bacterium that grows in acid conditions.

Acquired immunity—Resistance to antigens, developed during life, by producing antigen-specific antibodies.

Aerobe—A microbe that requires oxygen.

Aerosol—Tiny moisture particles expelled into the air.

Agar—A gelatinlike material used for growing bacteria and molds; a polysaccharide made by marine algae.

Airborne—Carried through the air on tiny particles or in moisture droplets.

Algae—A microbe that does photosynthesis but does not have cellular structure of plants.

Alkaliphile—A bacterium that grows in alkaline or basic conditions.

Anaerobe—A microbe that can grow without oxygen, or requires conditions lacking or very limited in oxygen.

Antibiotic—A natural or synthetic substance that kills microbes or inhibits their growth.

Antibiotic-resistant—Having the ability to live in the presence of an antibiotic.

Antigen—A compound on the surface of a cell that identifies it to the body as self or nonself.

Antigenic shift—A major change in the antigens on influenza viruses; it is the reason new flu vaccines must be made each year.

Antimicrobial—Killing or inhibiting microbes.

Antiseptic—A substance that removes microbes, usually on skin; refers also to microbe-free conditions.

Bacillus—A rod- or cigar-shaped bacterium. Plural: bacilli.

Bacteria—Single-celled microorganisms with a cell wall and no organelles. Singular: bacterium.

Bacteriophage—A virus that infects bacteria.

Biocide—A substance that kills living things.

Biofilm—A mixture of microbes and their excretions that adhere to surfaces in the presence of a liquid flow.

Bioremediation—The use of microbes to digest or neutralize toxins in the environment.

Biotechnology—The industry based on genetic engineering and devoted to using cells and cellular components to create new drugs and products.

Black mold—A mold that turns black when it grows, usually refers to *Stachybotrys*.

Bloom—A sudden growth of microbes to very large numbers.

cc—Cubic centimeter; equals one milliliter.

CDC—The Centers for Disease Control and Prevention.

Chemotherapeutic agent—A chemical or drug used to treat disease by killing specific cells; the drug used in chemotherapy.

Chromosome—A structure in a cell that carries all the cell's genes.

-cide—Suffix; kill.

Coccus—A round or spherical bacterium. Plural: cocci.

Coliforms—A diverse group of bacteria that ferment the sugar lactose, producing gas, and grow within forty-eight hours at 95 degrees F (35 C); used by water industry to indicate presence of fecal pollution.

Compound—A substance composed of two or more chemical elements.

Contact time—The minimum time required for a disinfectant or sanitizer to kill microbes.

Contagious—Describes a type of disease that is spread from person to person.

Contaminant—A microbe or chemical that spoils a substance or makes it unfit for use.

Contaminated—Containing an unwanted microbe.

Culture—To grow a microbe in a laboratory, or the resulting population of microbes produced after incubation.

Diatom—A type of algae that contains silica.

Dilution effect—Making a toxin harmless by mixing it with large volumes of water or other liquid.

Disease—A measurable condition in which a system, organs, or tissues do not function properly, affecting health and usually including symptoms.

Disinfectant—A substance that kills all microbes except bacterial spores.

DNA—Deoxyribonucleic acid; the genetic material in all self-reproducing cells and some viruses, a double-stranded molecule.

Domoic acid—A poison produced by certain diatoms.

Dormant—A condition of some microbes in which they are alive but metabolize very slowly and do not reproduce.

Drug—Any substance that when taken into the body can modify one or more functions.

Emerging—A new or new form of a disease in which incidence is increasing or expected to increase.

Enteric—Of the digestive tract.

Enzyme—A substance, usually a protein, that helps biological reactions progress.

EPA—U.S. Environmental Protection Agency.

Exotoxin—A toxin made by a microbe and excreted outside its cell.

Exponential—Changing at an increasingly high rate; same as logarithmic.

Extracellular—Outside the cell.

Extremophile—A microbe that lives in conditions far outside the norm, usually environments with extremes in temperature, dryness, acidity, salt, or pressure.

Facultative—Being able to grow with or without a specific condition, i.e., oxygen.

FDA—U.S. Food and Drug Administration.

Fecal—Made of or containing feces.

Filamentous—Having long extensions, used for growing through an environment.

Foodborne—Carried in food.

Fungus—A yeast, mold, or mushroom having cells containing organelles. Plural: fungi.

Gene—A segment of DNA that carries instructions for the cellular production of a substance.

Gene therapy—Treatment for disease by incorporating a gene(s) from outside the body or replacing a gene in the body's own DNA.

Gene transfer—The movement of a gene from one cell to another cell.

Genome—One complete copy of all of a cell's genetic material.

Genus—The first name in the most specific level of describing a living organism, i.e., *Staphylococcus* is the genus name for *Staphylococcus aureus*. Plural: genera.

Germ—Slang term for a microbe, usually used for harmful microbes.

Germicide—A substance that kills microbes.

Gram-negative—Bacteria that do not retain a purple stain in the Gram stain procedure; seen as pink under a microscope.

Gram-positive—Bacteria that retain a purple stain in the Gram stain procedure; seen as dark blue or purple under a microscope.

GRAS—Generally Recognized As Safe; a substance or food with historical evidence that it is safe for humans.

Halophile—A microbe that lives in high salt conditions.

Hepatitis—Inflammation of the liver, often caused by an infectious agent.

Herd immunity—A circumstance within a population in which a minimum percent of members are immune to an infection, making the infection's spread improbable.

Hygiene—The observance of cleaning practices for the body and sanitation for the surroundings to reduce the spread of infection.

Hypochlorite—A chemical unit containing chlorine and used in bleach as a disinfectant.

Immune system—The body's organs and cells that provide defense against microbial infection.

Immunity—Ability to use body organs and cells to resist a specific infectious agent.

Inanimate—Nonliving.

Incidence—The fraction of a population that contracts a disease in a given period of time.

Indicator bacteria—Bacteria that when detected in a sample of water are a sign of contamination by fecal bacteria.

Inert—Ingredient in an antimicrobial product that does not have activity against microbes.

Infection—Invasion or growth of microbes in the body.

Infectious agent—Any microbe that invades the body and initiates infection.

Infectious dose—The approximate minimum number of pathogen cells needed to start an infection.

Innate immunity—Resistance against nonspecific antigens that is present at birth.

Inoculate—Putting a small amount of microbes into growth media.

Ion—A negatively or positively charged atom.

Latency—A disease condition during which time there are no symptoms and the pathogen may be dormant.

Localized—An infection that does not spread across the skin or into the bloodstream.

Logarithmic—Same as exponential.

Lysis—Breaking apart of cells.

Media—The broth or agar formulas used to grow microbes. Singular: medium.

Mesophile—A microbe that grows in a moderate temperature range, 50 to 120 degrees F (10–50 C).

Metabolism—All the chemical and enzyme reactions of a cell or organism to generate energy and maintain life.

Microbial load—The total amount of microbes in a food or other substance.

Microenvironment—A location that provides unique conditions for microbial growth.

Microorganism—Same as microbe; a bacterium, mold spore, yeast, or protozoa cell.

Mildew—Slang term for visible mold that grows on surfaces in damp conditions.

Mold—Fuzzy form of fungus growth.

Molecule—A specific combination of atoms.

Mortality rate—The number of deaths in a population from a specific disease in a given period of time.

MRSA—Methicillin-Resistant *Staphylococcus aureus.*

Mucous membrane—The lining of body passages that are exposed to air, and usually containing mucus-secreting cells.

Mutation—A change in the correct gene sequence of DNA or the correct nucleotide sequence of a gene.

Mycotoxin—A poison excreted by a fungus.

Nanotechnology—The creation and use of substances or devices on a nanometer-length scale.

Native—Resident to the body or an environment.

Neurotoxin—A substance that interferes with nerve function.

Niche—A habitat to which a particular cell or organism adapts due to a specific trait.

Nucleotide—The basic unit of DNA, made of a nitrogen compound, a sugar, and a phosphorus unit.

Nutrient—A compound that supplies a cell with a portion of the total chemicals needed to generate energy and maintain life.

Obligate—Requiring a particular condition, i.e., an obligate anaerobe requires an environment without oxygen.

Opportunistic—Being normally harmless but able to infect if host susceptibility is weakened.

Organelle—A distinct membrane-enclosed section inside a cell that carries out specific processes; not found in bacteria.

Outbreak—A sudden increase in the incidence of a disease.

Parasite—An organism that derives nutrients by living on a host organism.

Pathogen—A microbe that causes disease in humans, animals, or plants.

PCR—Polymerase chain reaction, i.e., a method using the enzyme DNA polymerase to make large amounts of gene copies from a small amount of an original gene.

Phagocytosis—The enveloping and ingestion of particles or cells by other cells.

Plankton—Small organisms and microbes, often algae, suspended in marine waters.

Plasmid—A small circular piece of DNA separate from the chromosome, in bacteria.

Pneumococcus—General term usually referring to *Streptococcus pneumoniae.*

Preservative—A synthetic or natural substance that inhibits the growth of microbes in a product.

Propionibacteria—General term for skin bacteria of the *Propionibacterium* and *Corynebacterium* genera.

Protozoa—A single-celled microbe with internal structures and without a cell wall.

Psychrophile—A microbe that grows in cold temperature, usually below 60 degrees F (15 C).

Rare biosphere—A unique ecosystem that supports life found in few or no other places on earth.

Red tide—A bloom, or sudden growth, of marine algae with a reddish color.

Reemerging—A disease thought to be controlled in a population, but increasing in incidence rate, or expected to increase.

Reservoir—A continual source of an infectious agent.

Resident—A microbe that naturally lives on the body, or other environment; native.

Resistant—Able to withstand antimicrobial chemicals and drugs.

RNA—Ribonucleic acid, i.e., a single-stranded molecule necessary for DNA replication and protein synthesis.

Rod—Bacillus.

Sanitizer—A substance that reduces bacteria to safe levels, usually by reducing the bacteria 99.9 percent.

Septicemia—The presence of pathogens in the blood. Distinguished from sepsis, the spread of pathogens in blood and tissue causing inflammation and fever.

Serotype—A category of microbe within its species based on cellular structures or antigens.

Species—The last name in the most specific level of describing a living organism, i.e., *aureus* is the species name in *Staphylococcus aureus*.

Spore, bacterial—An almost indestructible, dormant form of a bacterial cell.

Spore, mold—A unicellular reproductive structure in some species.

Staphylococci—General term for round bacteria that grow in clusters.

Sterilizer—A substance that kills all microbes including bacterial spores; sporicide.

Strain—A unique form of a bacterial or protozoal species based on a genetic trait.

Streptococci—General term for round bacteria that grow in strands or chains.

Superbug—(a) A bacterium containing a nonnative gene(s) for carrying out specific intended reactions, or (b) a bacterial species that is resistant to antibiotics and/or chemicals.

Susceptibility—Lack of resistance to an infection or disease.

Thermophile—A microbe that grows in hot temperatures, optimum about 120 to 140 degrees F (50 to 60 C).

Toxic mold—Any mold that produces a mycotoxin or causes respiratory illness.

Toxin—A poison produced by a microbe.

Traceback—Process of determining the origin of an outbreak.

Transient—A microbe that does not naturally live on the body and does not remain there.

Transmission—The transfer of a pathogen from an infected source to a healthy person.

USDA—U.S. Department of Agriculture.

Virulence—The capacity of a pathogen to infect.

Virus—A submicroscopic particle containing minimal genetic material, which must infect a host cell to propagate.

Waterborne—Carried in water.

Xerophile—A microbe that lives in very dry conditions.

Yeast—A unicellular, or single-celled, fungus.

MICROBIOLOGY RESOURCES

When seeking information on microbes and microbiology, refer to publications from respected professional organizations, government agencies, or universities. If you still don't trust the message, go to similar scholarly resources for evidence of conflicting ideas. Do not depend on chat rooms, blogs, or magazines devoted to fashion, movie stars, home cures, and so on. Magazines such as *Scientific American, New Scientist, Science,* and *National Geographic* are reputable sources. University Web sites and those of world or national health organizations and professional microbiology associations are places to find factual information.

Additional reading
The Secret Life of Germs by Philip M. Tierno Jr., 2001.
Don't Touch That Doorknob! by Jack Brown, 2001.
The Germ Freak's Guide to Outwitting Colds and Flu by Allison Janse with Charles Gerba, 2005.
Where the Germs Are by Nicholas Bakalar, 2003.
My Office is Killing Me! by Jeffery C. May, 2006.
Carpet Monsters and Killer Spores by Nicholas P. Money, 2004.

BIBLIOGRAPHY

Several resources were used for this book. The resources that provided information for the majority of topics were (1) publications and data from the Centers for Disease Control and Prevention, the U.S. Food and Drug Administration, and the American Society for Microbiology; (2) *Microbiology, An Introduction,* 8th edition, by Tortora, Funke, and Case, 2004; (3) *Disinfection, Sterilization, and Preservation,* 4th edition, by Block, 1991; (4) *Environmental Microbiology* by Maier, Pepper, and Gerba, 2000; and (5) *Principles and Practices of Disinfection, Preservation, and Sterilization,* 3d edition, edited by Russell, Hugo, and Ayliffe. Resources specific to chapters are listed below.

Chapter 1
Environmental Literary Council.
University of California Museum of Paleontology.
Crick, F. Nobel lecture, December 11, 1962.
Dykhuizen, D. E. 1998. *Antonie van Leeuwenhoek* 73:25–33.
Lancefield, R. 1928. *Journal of Experimental Medicine* 47:91, 469, 481, 843, and 857.
Mullis, K. Patent No. 4,683,202, July 28, 1987.
Peltola, J., Andersson, M. A., Haahtela, T., Mussalo-Rauhamaa, H., Rainey, F. A., Kroppenstedt, R. M., Samson, R. A., and Salkinoja-Salonen, M. S. 2001. *Applied and Environmental Microbiology* 67:3269–3274.
Shelton, B. G., Kirkland, K. H., Flanders, W. D., and Morris, G. K. 2002. *Applied and Environmental Microbiology* 68:1743–1753.
Ward, B. B. 2002. *Proceedings of the National Academy of Sciences.*

Watson, J. Nobel lecture, December 11, 1962.
Wilkins, M. Nobel lecture, December 11, 1962.

Chapter 2
American Society for Microbiology.
Environmental Literacy Council.
Marine Biological Laboratory, Woods Hole, MA.
National Foundation for Infectious Diseases.
Society for General Microbiology.
Pure Water Handbook. 1991. Osmonics, Inc.
Gerba, C. University of Arizona, personal communication, 1998.
Sobsey, M. University of North Carolina, personal communication, 1997.
Smith, M., Bruhn, J., and Anderson, J. 1992. *Nature* 356:428.

Chapter 3
National Guideline Clearinghouse.
Agency for Healthcare Research and Quality.
American Scientific Laboratories, LLC.
American Society for Microbiology.
The Association for Science Education.
The Cleveland Clinic.
Dental Sciences, University of Newcastle upon Tyne.
United States Mint.
University of Minnesota Extension Service.
Wall Street Journal, January 6, 2006.
World Health Organization.
Ak, N. O., Cliver, D. O., and Kaspar, C. W. 1994. *Journal of Food Protection* 57:16 and 23.
Bart, K. J. (ed.). 1984. *Pediatric Infectious Disease Journal* 4:124.
Burge, H. A. 1990. *Toxicology and Industrial Health* 6:263.
Chamberlain, N. R. Microbiology at www.suite101.com, September 15, 2000.
Cliver, D., 1994, University of California–Davis, personal communication.
Dixon, B. *Microbe,* May 2005.
Gerba, C., *BBC News,* British Broadcasting Company, March 12, 2004.
Gerba, C., *CBS News,* Columbia Broadcasting System, September 28, 2000.
Gerba, C. 1998, University of Arizona, personal communication.
Ikawa, J. I., and Rossen, J. S. 1999. *Journal of Environmental Health,* July/August.

Kirchheimer, S. WebMD, www.medicinenet.com, July 12, 2004.

Klein, J. O. 1986. *Reviews of Infectious Diseases* 8:521.

Lillard, A. *Medical Laboratories,* University of Iowa, October 14, 1999.

Loeb, M., Craven, S., McGeer, A., Simor, A., Bradley, S., Low, D., Armstrong-Evans, M.,

Moss, L., and Walter, S. 2003. *American Journal of Epidemiology* 157:40.

Loeb, M., McGeer, A., McArthur, M., Peeling, R. W., Petric, M., and Simor, A. E. 2000. *Canadian Medical Association Journal* 162:1133.

Maibach, H. I., and Aly, R. 1981. *Skin Microbiology,* Springer-Verlag, New York.

Marston, W., *New York Times,* February 23, 1999.

Marques-Calvo, M. S. 2004. *Journal of Industrial Microbiology and Biotechnology* 31:255.

Meer, R. R., Gerba, C. P., and Enriquez, C. E. 1997. *Dairy Food and Environmental Sanitation* 17:352.

Nicolle, L. E., Strausbaugh, L. J., and Garibaldi, R. A. 1996. *Clinical Microbiology Reviews* 9:1.

Noble, W. C. 1981. *Microbiology of Human Skin,* Lloyd-Luke Ltd., London.

Osterwell, N. www.webmd.com, May 23, 2001.

Pence, A. *San Francisco Chronicle,* June 5, 2002.

Pitts, B., Stewart, P. S., McFeters, G. A., Hamilton, M. A., Willse, A., and Zelver, N. 1998. *Biofouling* 13:19.

Pitts, B., Willse, A., McFeters, G. A., Hamilton, M. A., Zelver, N., and Stewart, P. S. 2001. *Journal of Applied Microbiology* 91:110.

Robinson, T. CTW Features, *Santa Barbara News-Press,* May 28, 2006.

Rosenberg, M. *Scientific American,* April 2002.

Sankaridurg, P. R., Sharma, S., Willcox, M., Naduvilath, T. J., Sweeney, D. F., Holden, B. A., and Rao, G. N. 2000. *Journal of Clinical Microbiology* 38:4420.

Sattar, S. 1994, University of Ottawa, personal communication.

Sharkey, J. *New York Times,* March 11, 2001.

Simmons, R. B., Noble, J. A., Price, D. L., Crow, S. A., and Ahearn, D. G. 1997. *Journal of Industrial Microbiology and Biotechnology* 19:150.

Tierno, P., *Today Show,* National Broadcasting Company, March 25, 2005.

Tierno, P., www.wavy.com.

WHO. 2006. *Tuberculosis and Air Travel: Guidelines for Prevention and Control,* 2d ed.

Yarlott, N. *Opflow,* November 2000.

Young, F. E. *FDA Consumer,* January 2003.

Chapter 4

Access, American Water Works Association, August 2006.

American Meat Institute.

Association for Professionals in Infection Control and Epidemiology.

Center for Food Safety and Applied Nutrition, U.S. Food and Drug Administration.

Center for Global Food Issues, Hudson Institute.

Dairy Science and Technology, University of Guelph.

Department of Bacteriology, University of Wisconsin–Madison.

Department of Food Science and Human Nutrition, Washington State University.

Department of Health and Human Services.

ExtoxNet, University of California–Davis.

Minnesota Department of Health.

Morbidity and Mortality Weekly Report, The Centers for Disease Control and Prevention.

National Resources Defense Council.

The Ohio State University Extension.

Opflow, American Water Works Association, June 2006.

Physicians Committee for Responsible Medicine. *Foodborne Illness,* December 8, 2003.

Problem Organisms in Water: Identification and Treatment, 1995. American Water Works Association.

Tech Brief, National Environmental Sciences Center, Fall 2004, Spring 2006.

U.S. Environmental Protection Agency.

Virginia Cooperative Extension, Virginia Polytechnic and State University.

Banwart, G. J. 1989. *Basic Food Microbiology.* Chapman & Hall, New York.

Burnett, S. L., and Beuchat, L. R. 2000. *Journal of Industrial Microbiology and Biotechnology* 25:281.

Devine, D. and Jackson, M. *Tucson Weekly,* December 8, 2005.

Gibson, G. R., and Rastall, R. A. 2004. *American Society for Microbiology News* 70:224.

Korzeniewska, E., Filipkowska, Z., Domeradzka, S., and Wlodkowski, K. 2005. *Polish Journal of Microbiology* 54 Suppl: 27.

Logsdon, G. S., Schneider, O. D., and Budd, G. C. 2004. *Journal of the American Water Works Association* 96:7.

Mollenkamp, B. *Sanitary Maintenance,* August 2004.

Romans, J. R., and Ziegler, P. T. 1994. *The Meat We Eat.* Interstate Publishers, Danville, IL.

Scott, E., and Bloomfield, S. 1993. *Letters in Applied Microbiology* 16:173.

Zhang, J. *Wall Street Journal,* November 30, 2005.

Chapter 5

Alliance for the Prudent Use of Antibiotics.

Association for Assessment and Accreditation of Laboratory Animal Care International.

Center for Food Security and Public Health, Iowa State University.

Ciba Specialty Chemicals.

The Clorox Newsline, The Clorox Company.

Cosmetic, Toiletry, and Fragrance Association.

Consumer Specialty Products Association.

Minnesota Extension Service, University of Minnesota.

Akimitsu, N., Hamamoto, H., Inoue, R., Shoji, M., Akamine, A., Takemori, K., Hamasaki, N., and Sekimizu, K. 1999. *Antimicrobial Agents and Chemotherapy* 43:3042–3043.

Beck, W. C. 1984. *Association of Perioperative Nurses Journal* 40:172–176.

Chase, M. *Wall Street Journal,* January 20, 2006.

Chuanchuen, R., Beinlich, K., Hoang, T. T., Becher, A., Karkhoff-Schweizer, R. R., and Schweizer, H. P. 2001. *Antimicrobial Agents and Chemotherapy* 45:428–432.

DeNoon, D. 2001. WebMD Medical News Archive.

Entani, E., Asai, M., Tsujihata, S., Tsukamoto, Y., and Ohta, M. 1997. *Kanse Kasshi* 71:443–450.

Fraise, A. P. 2002. *Journal of Antimicrobial Chemotherapy* 49:11–12.

Gilbert, P., McBain, A .J., and Bloomfield, S. F. 2002. *Journal of Antimicrobial Chemotherapy* 50:137–139.

Glaser, A. 2004. *Pesticides and You* 24:12–17.

Gordon, S. *Healthscout,* July 25, 2000.

Juven, B. J., and Pierson, M. D. 1996. *Journal of Food Protection* 59:1233–1241.

Kiefer, R. J. 1998. *Chemical Times and Trends,* 21:48–50.

Kivanc, M., and Akgul, A. 1986. *Flavour and Fragrance Journal* 1:175–179.

Levy, S. *Scientific American,* March 1998.

Lewis, R. 1995. *FDA Consumer Magazine,* Vol. 29.

Lis-Balchin, M., and Deans, S. G. 1997. *Journal of Applied Microbiology* 82:759–762.

McMurry, L. M., Oethinger, M., and Levy, S. B. 1998. *Federation of European Microbiological Societies Microbiology Letters* 166:305–309.

Moken, M. C., McMurry, L. M., and Levy, S. B. 1997. *Antimicrobial Agents and Chemotherapy.* 41:2770–2772.

Neuman, C. 1998. *Chemical Times and Trends,* 21:35–48.

Parnes, C. A. 1997. *Chemical Times and Trends* 20:34–43.

Rusin, P., Orosz-Coughlin, P., and Gerba, C. 1998. *Journal of Applied Microbiology.* 85:819–828.

Silver, S., Phung, L. T., and Silver, G. 2006. *Journal of Industrial Microbiology and Biotechnology* 33:627–634.

Smith-Palmer, A., Stewart, J., and Fyfe, L. 1998. *Letters in Applied Microbiology* 26:118–122.

Suller, M. T. E., and Russell, A. D. 2000. *Journal of Antimicrobial Chemotherapy* 46:11–18.

Ulene, V. *Los Angeles Times,* February 6, 2006.

Wendorff, W. L., and Wee, C. 1997. *Journal of Food Protection* 60:153–156.

Chapter 6

Commoncold, Inc.

Immunization Action Coalition.

"Influenza 1918" Public Broadcasting System.

Karolinska Institutet.

Medical News Today, September 29, 2006.

National Immunization Program, CDC.

National Institute of Allergy and Infectious Disease.

National Nosocomial Infection Surveillance System, CDC.

National Vaccine Information Center.

University of Maryland Medical Center.

World Health Organization.

Burke, L. *George Mason University Alumni Magazine,* Winter 2003.

Enserink, M. *ScienceNow,* February 6, 2004, American Association for the Advancement of Science.

Gorman, C. *Time Magazine,* June 26, 2006.

Rasmussen, C. *Los Angeles Times,* November 20, 2005; March 5, 2006.

Rutala, W. A., and Weber, D. J. 1997. *Infection Control and Hospital Epidemiology.* 18:609.

Sattar, S., Jacobsen, H., Springthorpe, V. S., Cusack, T. M., and Rubino, J. R. 1993. *Applied and Environmental Microbiology* 59:1579.

Tansey, B. *San Francisco Chronicle,* December 1, 2006.

Walker, C. *National Geographic News,* March 10, 2004.

Chapter 7

American Academy of Dermatology.

AVERT, United Kingdom.

California Department of Health Services.

Center for Biolfilm Engineering, Montana State University.

The Chlorine Chemistry Council.

Dermatology Insights, fall 2001.

Department of Health and Human Services News, October 31, 1996.

Directors of Health Promotion and Education.

Eurekalert Science News, November 27, 2006, American Association for the Advancement of Science.

Faculty of Medicine, University of Manitoba.

Global Census of Marine Life.

Infectious Diseases Society of America, www.legionella.org.

Karolinska Institutet.

Marine Biological Laboratory, Woods Hole, MA.

National Institute of Allergy and Infectious Disease.

National Oceanic and Atmospheric Administration.

National Swimming Pool Foundation.

New Jersey Medical School Global Tuberculosis Institute.

School of Biological Sciences, University of Leicester.

Sportsmedicine.about.com, New York Times Company.

University of Texas–Houston Medical School.

World Health Organization.

Barquet, N. and Domingo, P. 1997. *Annals of Internal Medicine* 127:635.

Breslin, M. M. *Chicago Tribune,* October 29, 2006.

Cromley, J. *Los Angeles Times,* June 6, 2006.

Halsey, E. Cable News Network, November 1, 1996.

Hope, J. *Daily Mail,* January 3, 2005.

Louria, D. B. 1998. In *Emerging Infections,* Scheld W. M., and Hughes, J. M. (eds), American Society for Microbiology.

McFadden, R. D. *New York Times,* December 5, 2006.

Parry, C., and Davies, P. D. O. 1996. *Journal of Applied Bacteriology* 81:23S.

Read, M. *Associated Press,* October 18, 2006.

Russell, S. *San Francisco Chronicle,* October 20, 2006.

Schlossberg, D. (ed). 2004, *Infections of Leisure,* American Society for Microbiology.

Stobbe, M. Associated Press, June 23, 2006.

Washburn, J., Jacobsen, J. A., Marston, E., and Thorsen, B. 1976. *Journal of the American Medical Association* 235:2205.

Woolhouse, M. E. J. *Microbe,* November 2006.

Chapter 8

Bob's Red Mill Natural Foods.

Graduate College of Marine Studies, University of Delaware.

Human Genome Project.

Marine Biological Laboratory, Woods Hole, MA.

The Natural History Museum, London.

Oak Ridge National Laboratory.

Seattle Times, November 19, 2004.

Washington State Patrol Crime Laboratory.

www.careerbuilder.com.

Baross, J. A., and Deming, J. W. 1983. *Nature* 303:423.

Brock, T. D., and Freeze, H. 1969. *Journal of Bacteriology* 98:289.

Deming, J. W. *2nd International Colloquium of Marine Bacteriology* October 1984.

Fisher, B. A. J. 2004. *Techniques of Crime Scene Investigation,* CRC Press, Boca Raton, FL.

Perlman, D. *San Francisco Chronicle,* October 20, 2006.

Sogin, M. L., Morrison, H. G., Huber, G. S., Welch, D. M., Huse, S. M., Neal, P. R., Arrieta, J. M., and Herndl, G. J. 2006. *Proceedings of the National Academy of Sciences* 103:12115.

ACKNOWLEDGMENTS

I would like to thank a number of people for their patient attention to the science presented in this book. Appreciation goes to Dr. Carlos Enriquez, Dr. Dana Gonzales, Dr. Kristina Mena, Carole Parnes, MS, and Dr. Robert Ruskin for their keen attention to technical details. I thank Bonnie DeClark, Priscilla Royal, Sheldon Siegel, Meg Stiefvater, and Janet Wallace for their insightful thoughts and questions.

I greatly appreciate the valuable help received from Keith Wallman, Peter Jacoby, and Jennifer Kasius, who guided the progress of this book from rough manuscript to finished product. Finally, but not least, I thank Jodie Rhodes, who sees opportunity and promise around every corner.

INDEX

ABOUT THE AUTHOR

Anne E. Maczulak, PhD, has been a microbiologist in industry and academia. She is currently a registered quality assurance consultant for biotechnology, pharmaceutical, consumer product, and chemical companies.

Anne's career has covered diverse topics in microbiology. She has studied the protozoa and bacteria of horse and cow digestive tracts and is one of a relatively small group of microbiologists who were trained in the Hungate method of growing anaerobic—without oxygen—microbes. In industry, she conducted studies on odor-causing skin bacteria and the scalp yeasts that promote dandruff. She has participated in dermatology research and clinical tests on wound-healing medications, antimicrobial soaps, and foot fungi treatments. In the consumer products industry, she developed biological treatments for clogged drains, disinfectants for homes and hospitals, and water purifiers.

Throughout her career, Anne has researched a range of microbiological environments and specialized microorganisms. Some of these include the anaerobic populations of the cecum and rumen, cold-tolerant enzymes from soil bacteria, biofilms, methane-producing bacteria, spore-forming bacteria, household molds and fungi, viruses, and protozoan parasites found in natural waters. She has collaborated with several academic teams on projects: biofilm growth at the Center for Biofilm Engineering at Montana State University, cold-tolerant enzyme production in soil bacteria at Pennsylvania State University, biodegradative enzyme production in water bacteria at the Center of Marine Biotechnology of the University of Maryland, and household microbes at Georgia State University and the University of South Florida.

During her years as a bench microbiologist, Anne used an assortment of laboratory techniques, including DNA isolation and hybridization, large-volume fermentation cultures, PCR, gel electrophoresis, cloning, and monoclonal antibodies. She has developed a variety of test methods and specialized culture media.

Dr. Maczulak has given presentations on microbiology at the American Society for Microbiology, the American Academy of Dermatology, the Environmental Health Association, and the Society of Quality Assurance. She has authored several journal articles and has taught university classes.

In addition to her PhD from the University of Kentucky, Anne holds an MBA from Golden Gate University in San Francisco. She completed postdoctoral research at the New York State Department of Health in Albany, and her master's and undergraduate studies at The Ohio State University.

Anne offers a unique blend of basic research and practical knowledge. Her ability to see the "big picture" and to deliver technical subjects in an easy-to-understand manner enables her to teach nonscientists about the mysterious world seen under a microscope.